Mobile Marketing

for Your

Local Business

Co-Authored

By

Mario Brown, Brian Anderson

Irina Finkler, Susan Banks

Jack Hopman, and Jeff Cepuran

Mario Brown, Brian Anderson, Irina Finkler, Susan Banks, Jack Hopman, and Jeff Cepuran

Contents

Forward by
Mario Brown & Brian Anderson

Mobile marketing is transforming the way businesses of all sizes interact with their clients. Whether a local business or a global 1,000 enterprise, mobile marketing is an integral part of "personalized" marketing. Unfortunately, there have been relatively few books on mobile marketing specifically for the small business. Most of the mobile marketing strategies and content that I see are predominately based on theories, and not "real world" proven strategies and experiences. Mobile Marketing For Your Local Business puts an end to that pattern.

Mobile Marketing For Your Local Business immediately delivers real world examples and content any business owner can use – from the very beginning. After a solid high level introduction to the concepts of mobile marketing it takes us one chapter at a time through key mobile marketing methods that will both add new clients and increase retention and relationship with existing clients. This book covers SMS text message marketing, QR codes, mobile Apps, as well as, Mobile websites. Unlike many who gloss over these topics the contributing authors of Mobile Marketing For Your Local Business continue to deliver value for the reader with tips, strategies, examples, check lists, the ambitious small business owner has a virtual playbook at their disposal!

I have seen many examples of small businesses that have successfully leveraged mobile marketing to get more customers and increase revenues. There are many restaurant owners who turned around slow days and nights, through SMS loyalty marketing. Retail stores who captured new clients through mobile websites. Innovative small businesses using Apps to maintain a relationship with their most cutting edge clients. Automotive dealerships who increased the number of vehicles in their service department within days using coupons and text message marketing!

It is solutions and strategies like this one from a very smart business owner that will continue to allow some businesses to separate from their competition. Mobile is just that – a separator between the "winners" and "losers" in the war for customer's time and money. Harness the power of mobile marketing successfully for your business, and guarantee yourself visibility and customer loyalty that you alone control. No Google. No expensive television or radio. No outdated print media.

As a fellow mobile marketer who creates and innovates marketing solutions for local businesses using the viral aspects of mobile and social media I feel like Mobile Marketing For Your Local Business is a remarkable resource for all small business owners who want to capitalize on this constantly changing mobile marketing phenomena.

Introduction

Okay, let's face it! Mobile devices are here to stay!

Cell (or mobile) phones have gone from the brick-sized, shoulder-bag version that weighed as much as -- well, a brick -- to the stylish, chic version that fits in any pocket and goes with you everywhere. If you are like 85-90% of mobile device owners, you literally won't leave home without it! In fact, you'll turn around and go back home for it. Mobile phones are everywhere!

They have become so necessary to our lives that we sometimes find that we have both a personal cell phone and a business cell phone. We give them to our children, our employees, our grandparents and we are shocked when we discover someone under the age of 100 that doesn't own at least one.

Mobile Marketing, a new marketing trend, is the result of the growth in the use of mobile phones and other mobile devices, such as tablets & eReaders.

No business can afford to ignore these millions of screens that we all hold inches from our faces all day long. Done well, Mobile Marketing can be *the thing* that takes your success to the next level.

Who is This Book For?

This book has been written for the business owner who is ready to take advantage of this new marketing trend. A trend that is poised to take 2014 by storm! Mobile marketing is for the business owner who wants to be ahead of the trend, in front of the pack - instead of trailing it long after it's been around for years and used successfully by others.

If you own a real estate office, salon, car dealership, coffee shop, auto repair shop, restaurant or any other local business, you know that a current, regular paying customer is worth more to you than a customer you have to go out and find.

Once you have your paying customers, you may not have any way of keeping track of them – who they are, how many of them you have, what they buy when they come in to your shop, how often they buy and so you haven't been able to market directly to them – bring them back in your shop when you are having a sale on something they like/need/want and are willing to spend more money on.

Do you do your marketing the old-fashioned shotgun way? Blast the market with advertising in the direction of where you "think" is the right customer is? If you are hoping that someone who needs your product or service sees your ad and will simply come in and spend money, then you are never actually sure who your advertising reaches or how successful your campaigns are.

Mobile marketing can be a highly profitable way to communicate with your current customers and a very smart way to not only find new customers, but keep them all coming back to your business for many years to come. It's far easier to keep a customer than to get a new one!

If you are new to mobile marketing and simply want to learn more about it, or have already finished your research and want to know your next step – this book is for you.

Eye-Opening Mobile Marketing Statistics

Let's go over some recent statistics that I find not only mind-boggling, but eye opening as well – see if you agree with me after reading through these that mobile marketing is well worth your investing some time in.

They say a picture is worth a thousand words. All over the internet and in magazines and periodicals you can find statistics and graphs such as this one:

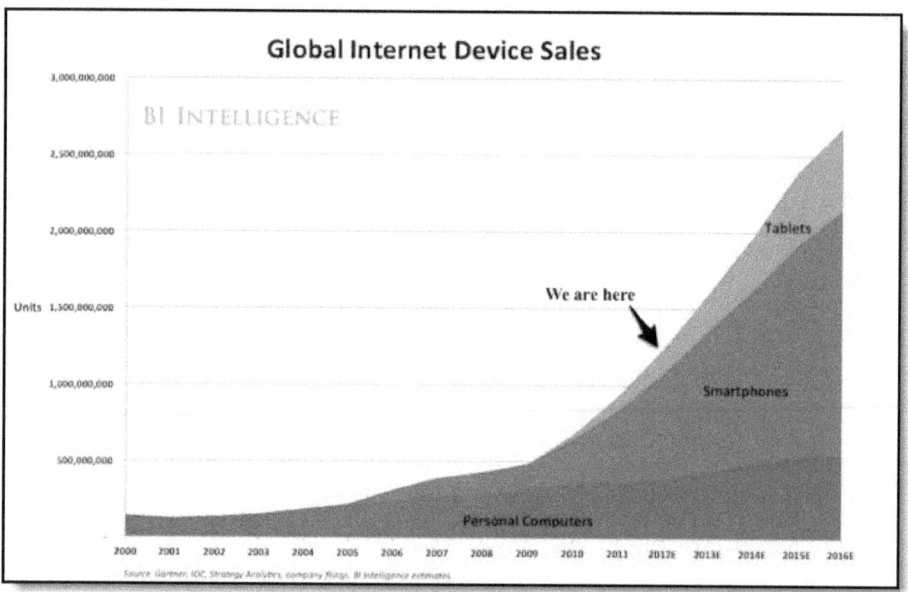

What this shows is that the growth of mobile sales is a fact of everyday life – as is the projection that this growth will soar even more steeply within the next four years. SnapHop quotes four statistical research companies as predicting: "Global internet usage will more than double by 2015, and most of these users will be mobile."

State of the Mobile World

- There are 7 billion people on Earth. 5.1 billion own a cell phone. 4.2 billion own a toothbrush. (Mobile Marketing Association Asia, 2011)

- 91% of all smart phone users have their phone within arm's reach 24/7 – (Morgan Stanley, 2012)

- It takes 26 hours for the average person to report a lost wallet. It takes 68 minutes for them to report a lost phone. (Unisys, 2012)

- Mobile device sales rose in 2011, with smartphones showing the strongest growth. Nokia remains the number one handset manufacturer, but Samsung is the leading smartphone hardware vendor. Android is now the top smartphone operating system.

- Powering more than 250 million devices, the Android OS runs on half of all smartphone shipped, with a user base increasing by 700,000 subscribers each day.

- Adults spend more media time on their mobile device than newspapers and magazines combined (eMarketer December 2011 - http://www.emarketer.com/Article.aspx?R=1008728)

Mobile Optimization

- 74% of consumers will only wait 5 seconds for a web page to load on their mobile device before abandoning the site.

- 46% of consumers are unlikely to return to a mobile site if it didn't work properly during their last visit.

- 71% of mobile browsers expect web pages to load almost as quickly or faster as web pages on their desktop computers.

- Source for above:

http://www.compuware.com/d/release/592528/new-study-reveals-the-mobile-web-disappoints-global-consumers

Mobile Search

- 70% of all mobile searches result in action within 1 hour. 70% of online searches result in action in one month. (Mobile Marketer, 2012)

- 61% of local searches on a mobile phone result in a phone call. (Google, 2012)

- 52% of all mobile ads result in a phone call. (xAd, 2012)

- 20% of telecom, 30% of restaurant, and 25% of movie searches are mobile (Google)

- Mentioning a location in mobile ads and search results can increase click-through rates up to 200%. (Source: http://www.mediapost.com/publications/article/171106/localized-creative-improves-click-through-rates-e.html)

Mobile Commerce

- Mobile coupons get 10 times the redemption rate of traditional coupons. (Mobile Marketer, 2012)

- Only 19% of US retailers have a mobile app. (Source: http://www.emarketer.com/Article.aspx?id=1008792&R=1008792)

- 24% of US tablet owners use their tablets to shop 2-3 times per month; 20% use them to shop more than once per week; and 12% use them to shop every day! (Source: http://www.emarketer.com/Article.aspx?id=1008792&R=1008792)

- Mobile marketing will account for 15.2% of global online ad spend by 2016. (Berg Insight, 2012)

- 1 in 5 smartphone users scan product barcodes, and nearly 1 in 8 compare prices on their phone while in a store. (Source: http://www.comscore.com/Press_Events/Presentations_Whitepapers/2012/2012_Mobile_Future_in_Focus)

- 39% of store walkouts, where consumers leave without buying, were influenced by smartphones. (Source: http://www.mobilecommercedaily.com/2011/01/13/87pc-of-retailers-agree-shoppers-can-find-better-deals-via-mobile-survey)

Mobile Social Media

- 71% of smartphone users that see a captivating TV, press or mobile advertisement will immediately do a mobile search.

- 91% of mobile internet access is for social activities, versus just 79% on desktops.

  (Source:

  http://tag.microsoft.com/community/blog/t/the_growth_of_mobile_marketing_and_tagging.aspx)

- 44% of Facebook's 900 million monthly users access Facebook on their phones. These people are twice as active on Facebook as non-mobile users (Facebook, 2012)

- QR code scans increased 300% in 2011 compared to 2010. (Source: http://www.scanlife.com/blog/2012/02/2011-finishes-strong-with-millions-of-new-users/)

- QR code (http://hop.mx/) usage jumped 617% from January to December 2011 in top 100 magazines (Nellymoser) http://www.prweb.com/releases/2012/1/prweb9140465.htm

Mobile Email/App/Text

- It takes 90 **minutes** for the average person to respond to an email. It takes 90 **seconds** for the average person to respond to a text message. (CTIA.org, 2011)

- If all US mobile internet time was condensed into an hour, 25 minutes of it would be spent on email. (Source: http://econsultancy.com/us/blog/9789-mobile-email-stats-infographic)

- From April to September in 2011, mobile email opens increased 34%, while webmail and PC opens decreased by 11% and 9.5%, respectively.

  (Source:

  http://www.returnpath.net/downloads/reports/returnpath_mobilemessagingwhitepaper.pdf)

- iPad users are loving it for email -- there's been a 73% increase in opens on those skinny little things.

  (Source:

  http://www.returnpath.net/downloads/reports/returnpath_mobilemessagingwhitepaper.pdf)

- Mobile email readership is at its peak on Saturday and at its lowest on Monday.

  (Source: http://infographiclist.com/2012/04/23/mobile-email-marketing-infographic/)

- 8 trillion text messages were sent in 2011

- Over 300,000 apps have been developed in the past 3 years. Apps have been downloaded 10.9 billion times.

- 64% of mobile phone time is spent on apps (http://na.ad-tech.com/sf/wp-content/uploads/DigitalConsumer.pdf)

- 13.4% texting, 11.1% browser, 5.5% social networking apps, 5.4% dialer, 5.3% email/IM, 2.3% music/video, 1.1% camera, 55.8% other apps

  (http://na.ad-tech.com/sf/wp-content/uploads/DigitalConsumer.pdf)

2013-Canada-Digital-Future+in+Focus.pdf (Comscore – March 2013)

Smartphone Market Penetration in Canada by Percent of Mobile Subscribers increased from 45% in December 2011 to 62% in December 2012 (and 33% in 2010)

US-Digital-Future-in-Focus-2013.pdf (February 2013)

page 33: "Smartphones continued to drive the mobile landscape in 2012, finally reaching 50-percent market penetration at the end of Q3.

Google and Neilsen just released a report in March of 2013 that was very interesting

Google_Neilson_mobile-search_March_2013.pdf

77% of mobile searches are done at home or work; 17% on the go – so people are searching on their phone even when they have a computer close by – because it is quicker and easier.

What is Mobile Marketing?

According to Wikipedia:

"Mobile marketing is marketing on or with a mobile device, such as a cell phone."... "Using interactive wireless media to provide customers with time and location sensitive, personalized information that promotes goods, services and ideas..."

Why Mobile Marketing is Important

Mobile is now considered the 7th mass media – the other 6 being Print, Recordings, Movies, Radio, Television, and Internet. They are referred to as mass media because they reach out and touch massive numbers of potential buyers at one time – think Superbowl ad.

What's So Special ...

About the mobile industry -- the newest mass media? It is the first mass media that is capable of doing everything the other six can do. We can read newspapers, watch movies, listen to music, read books, watch TV shows and we can surf the web – all on a smartphone.

More importantly, mobile phones have benefits that the other 6 media don't have.

- It is the only media that is truly personal and carried constantly. Mobile phones are not shared with others – even our spouses! Everyone has their own

- Mobile is the only media where each consumer can be identified individually (by their phone number) with a degree of accuracy not available with any other media

- Mobile is the first and only media that is always on. Messages, news, and promotions are delivered instantly and can be accessed 24/7 by consumers

- The mobile phone is the only media that is constantly with the owner

- Mobile phones have built-in payment mechanisms. Click to Buy is becoming more and more common. You can't run up to your television screen to buy that shirt just advertised or touch a button on your car radio to reserve those theatre tickets being advertised.

And the most incredible benefit of mobile is all this can be done without waiting – at that instant – talk about impulse buying. It's instant gratification at its finest.

- No need to power up the computer

- No need to download the picture from the camera

- No need to turn on the player and insert the DVD

- No printing the coupon, cutting it out and walking it to the store

- No waiting to go to the store to buy the book you want right now

Thanks to the "Click to Buy" feature, mobile users can have it NOW!

Today's Hidden Market

Mobile marketing is simply using mobile technology to help you reach your customers on a more personal and convenient level. It allows you to market more effectively by creating marketing campaigns that your customers or clients will enjoy and find helpful via their mobile devices.

Mobile also enables you to reach a "hidden" segment of target customers – ones who don't physically explore the town or use the Yellow Pages to find you and those who don't use personal computers. Instead, this hidden segment almost exclusively use their mobile devices – cell phones, Smartphones, iPads or other Android tablets.

This hidden segment is the one most likely to be _motivated to buy_. They are highly focused and live via their mobiles because they are used to immediate results. You won't have the usual seven to twelve visits to your store the average customer makes before buying – if they find you and you've got what they want, they buy straight away.

Our own society has progressed to a point where people are overloaded, stressed out and fractured psychologically. The result is a society that feels it needs (and is owed):

- Instant gratification

- Immediate results

- To feel connected in real-time

This is a society that finds interactivity gives the illusion of fulfilling all these needs. And the only place to get it in retail or business is by using their mobiles.

To reach them, you need to have a mobile marketing plan in place.

Psychology of an Average Local Mobile Consumer

The key personality traits of the average local mobile consumer is that they are ready to buy and looking for quick information right at their fingertips – and mobiles fill this need with a satisfying degree of instant gratification.

Let's peek under the hood just a bit further …

Mobile users tend to be obsessive-compulsive about their devices. We all know someone who always has his iPhone up to his ear (usually in the middle of real-person conversations – with us). Or the teenagers whose heads are always down, thumbs in motion, as they text at the speed of light. While the University of Arkansas states that a worrying number of mobile users do suffer from actual obsessive-compulsive disorder when it comes to their devices, one should be aware of the two, twin motivations behind obsessive cell use: Both pleasure... and a stress response.

"Evidence of compulsive behavior brings to light the notion that the underlying motivation to use a mobile phone is not pleasure, as predicted by addictions studies, but rather a response to heightened stress and anxiety," said Moez Limayem, one of the wheels behind the University of Arkansas study, which was to determine the potential dangers of mobile phone usage while driving. Zachary Steelman, Amr Soror, Moez Limayem, and Dan Worrell, "Obsessive Compulsive Tendencies as Predictors of Dangerous Mobile Phone Usage" (July 29, 2012). AMCIS 2012 Proceedings.

Whatever their reasons, those teenagers we see always texting are tomorrow's customers. That's one reason why experts can predict so confidently that mobile usage will double within the next four years.

Even older people have jumped on the bandwagon and learned to interact with social networks such as Facebook. And many have learned how to text – many of them grudgingly. Some see it as the best (and in a few cases, the only) way they can communicate with their children.

Interview with an Octogenarian

"It seems it's the only way I can ever connect with my grandchildren and great-grandchildren," complained Gwen Foster, 82, of Renfrew, Ontario. "They don't answer letters and even emails get ignored. I have the choice of being out of touch or going where I can find them."

Adds Foster: "It's not rudeness, much as it may seem like that to someone of my generation. It's simply that if you don't text and Facebook, you're not even on the same planet. You're simply out of their view."

Foster is one of a minority: She chooses not to interact via Facebook and complains that texting is "beyond" her – but she and her husband do own a mobile phone "for emergencies," and adds that "it's surprising how often it comes in useful when you're out and about." She uses two Apps, both of which were programmed into her phone by her daughter.

When asked why she thinks the younger generations don't respond well to other forms of communication, Foster rejected the common explanation offered by some of her peers. That it's a "lack of manners."

"I think cell phones just make it so much easier for them to communicate," she finally decides.

Know Your Customers

Understanding all these different personality types and varying motivations is part and parcel of accurately predicting how your particular, unique group of customers will react to mobile marketing.

You should do your own research. Whether it's asking customers who come in your business what they think about mobile marketing options: text messaging, QR codes, mobile search and/or a mobile app. Offer a QR code for them to scan that takes them to a quick survey online. Give them a simple questionnaire they can fill out by hand and give back to you. Or just have your employees ask a couple of questions – and keep track of the answers.

Be sure to offer those customers who take the time to answer your questions a gift for filling out the survey – 20% off their next purchase or free dessert with their next meal, etc. And don't forget to ask them to sign up for email, mobile alerts, mobile coupons or text messages.

Remember that study performed by The University of Arkansas? They kept track of the demographic data among their respondents:

- Average age was 28

- 66.6% male

- 80% had a college degree

- 75% were employed

- 57% were single

- 57.6% used a Smartphone over a standard mobile phone

This gives a very clear picture of a specific group. This is the type of information that, as a business owner, is so invaluable for you to know. If the majority of your customers are like our Octogenarian from earlier– over the age of 80, married, with a simple cell phone used "only for emergencies," then mobile marketing campaigns would most likely be a waste of your time and money. This demographic would respond better to direct-mail campaigns in large-print format.

It's really all about your customer, how well you know their habits and how to use that information to its best advertising advantage.

Demographic Research Online

There are two easy methods of conducting your own demographic research online.

Google It!

Google is your friend! It's amazing what you can research and discover just by asking a simple question in your Google search bar.

Just keep an eye open for the dates that the articles were published (see the red arrow in the screenshot below). Choose the newest results – they don't always show up at the top.

Google mobile phone usage by baby boomers

Web   Images   Maps   Shopping   More ▾   Search tools

About 2,700,000 results (0.45 seconds)

How marketers can use mobile technology to reach ba...
www.mobilemarketer.com › Opinion
Jan 11, 2011 – How marketers can use mobile technology to
reach baby boomers ... Most current studies show that while
most boomers have mobile phones, ...

How Baby Boomers Are Embracing Digital Media
mashable.com/2011/04/06/baby-boomers-digital-media/
Apr 6, 2011 – This year, some of the nearly 80 million Baby
Boomers in the United States ... It will challenge us to rethink
how we use the web and how we engage ... who have given up
landlines for mobile phones sparked 138 comments.
Pete Cashmore +1'd this

Why US Baby Boomers are Slow to Buy Smart Phone...
bigthink.com/.../why-us-baby-boomers-are-slow-to-buy-smart...
Apr 13, 2011 – Why US Baby Boomers are Slow to Buy Smart
Phones: ... In Japan, mobile media use among people 45 to 60
years old – Japan's Dankai ...

Baby Boomers Embracing Mobile Technology
www.marketingcharts.com/.../baby-boomers-
embracing-mobil...
Sep 24, 2007 – Baby Boomers Embracing Mobile Technology
... on their phone and one in four (25%) older mobile users
recall seeing ads on their phone.

78% of US Baby Boomers Are Online: Are Brands Rea...
therealtimereport.com › Blog
Apr 5, 2011 – 86.9% of Baby Boomers will have a mobile
phone in 2011 and in 2015 ... 40% of boomer mobil...

Direct Audience Measurement

You can also take the website addresses of your competitors and type them into an audience measurement analysis site such as Quantcast or Alexa.

Quantcast, for example, is a free direct audience measurement for all website owners that includes traffic (website visitors), demographics, business, lifestyle, interests and more. These sites not only show you a graph indicating whether or not your competitor's business is growing or declining, if you scroll down you will see instant general demographic results that can give you an idea whether or not their market is actually the same as, or similar to, your unique market...

For example, take Quantcast's analysis of The Olive Garden:

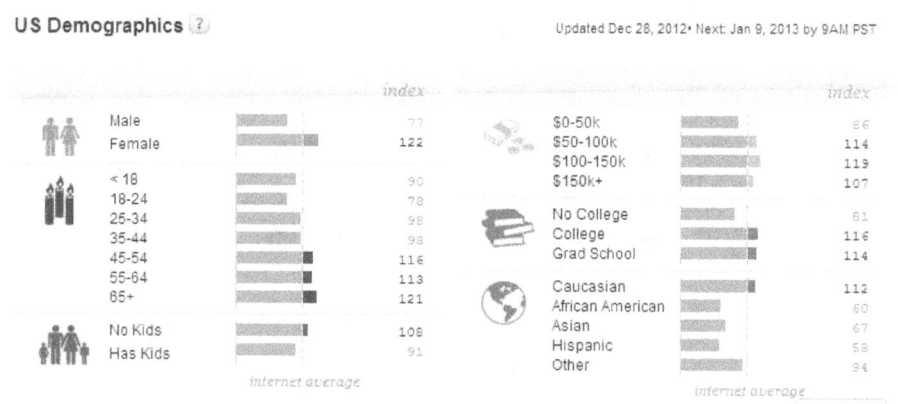

From this you can see that the strongest demographic group is females in the 45-64 age range, with children, earning between $100-$150k annual household income with a college degree. Caucasian women are more prone to eat at Olive Garden, but most races are fairly well represented, with Hispanic women the least likely to visit.

Getting sold on the idea of researching your target market online yet?

Combine the specific demographic information you learn about your target group from competitors against mobile usage statistics, such as the "mobile phone usage by baby boomers" Google searches in our example above and you should have a pretty good idea of whether or not mobile marketing is right for your particular audience.

Once you add your in-person questioning and/or questionnaires to the mix you will be able to eliminate a large portion of the errors that always occurs with purely statistical research.

Best of all, as a small business owner, you won't have to shell out thousands of dollars for research to some marketing company: This type of research gathered online and in person is free.

Components of a Local Mobile Campaign

Our Octogenarian, Gwen Foster, really hits the nail on the head with her comment about how easy mobile devices make communication. And that is why mobile devices are ideal as marketing tools for small businesses – especially local ones.

You've identified your target group (and their cell phone habits). At the end of this book, you'll find a worksheet offering suggested questions to ask on your in-store or in-office ballot. Feel free to customize questions to fit your unique customer base.

Be sure to replace the text in brackets, [Your Free Incentive] and [Your Business] with your unique data!

The components of a successful local mobile campaign can be broken down into four segments and we'll be going over these in the next chapters:

- Mobile Websites

- Text Messaging Marketing

- QR Code Marketing

- Mobile Apps

So, let's get started!

Meet Co-Author Mario Brown

Mario Brown grew up in Germany where he played professional Basketball and successfully studied foreign trade in Germany. He lived in Ecuador for 7 months in 2007 where he met the love of his life before moving to Miami in 2008. Mario recently returned to Ecuador for a holiday wedding where he married the love of his life. Together they moved back to Florida where Mario runs his businesses full-time.

In Miami he developed his passion for entrepreneurship and self-improvement and he now owns 2 highly successful online businesses.

All of his clients' websites rank on the first page of Google (without exception) and his own SEO Consultancy websites rank on top of the first page for the most competitive keywords - leaving his competition far behind him.

His company achieves these outstanding results by following basic but powerful Search Engine Optimization strategies.

Mario Brown is one of the most in-demand speakers on Online Marketing, Mobile & SEO Solutions, Leadership and High Performance in the world.

Mobile Websites

We've determined that mobile phones are a big part of most people's lives and have become a "necessity" and "can't-live-without" device for just about everyone; especially now that these phones are becoming smarter every day. Mobile phones are being used to access the internet and perform searches to find business information, phone numbers, directions and maps to a business's location. While a mobile or smartphone can do a variety of other amazing things, when a person is looking at a mobile website for a local business what they really want to know is:

- Where the business is located

- How do I get there

- Are they open right now

- What is their phone number

This makes having a mobile-friendly website a necessity. These are websites that are designed to make it possible for mobile users to access your critical business information without problems.

How Mobile Websites Increase Profits

More and more people are using mobile phones to get on the internet. If a potential customer is out on the road and trying to find a way to get to your store or call you then you need to have a mobile-friendly website.

Believe me, if your potential customer can't find your information easily then they will find your competitor's store and visit them instead. We are a society of instant gratification, so every business should use strategies to ensure their website is visible to everyone regardless of the means used to access the internet.

The following mobile websites statistics confirm exactly why your website must be mobile-friendly:

- By the year 2014, it is expected that more people will access the internet using a mobile device than using computers

- 71% of mobile users expect a website to load faster on their phones than it does on their computer

- 57% of mobile users will not recommend any business that has a mobile unfriendly website

- 40% of mobile users will move to competitor websites if they have a slight problem accessing a website

- 95% of mobile users will search for local businesses using their devices and 61% of these will call the business they find

Mobile Website Design

The reason behind the design of mobile websites is that mobile phones are much slower when it comes to processing power than a laptop or desktop computer. They also have smaller screens, so mobile-friendly websites are built specifically for viewing on mobile devices.

If you have ever tried accessing any website that is not mobile-friendly on your mobile phone, then you have experienced the frustration shared by millions of mobile device surfers worldwide. First of all, the website might take ages to bring up the home page and when it does, visitors will probably on see parts of the page and tiny, tiny print. This means that visitors will need to zoom in and out and scroll side to side in order to view the page.

For example, take Olive Garden's website. This is what you see, if you go to http://www.olivegarden.com/ on your personal computer:

Do you see how nearly impossible it is to read even when displayed as intended on a **full sized screen** on a desktop computer?

Now imagine this layout on a mobile device with a screen 1/10th the size of your average computer monitor.

However, here is what you would actually see if you entered the Olive Garden's website address _**on your mobile phone:**_

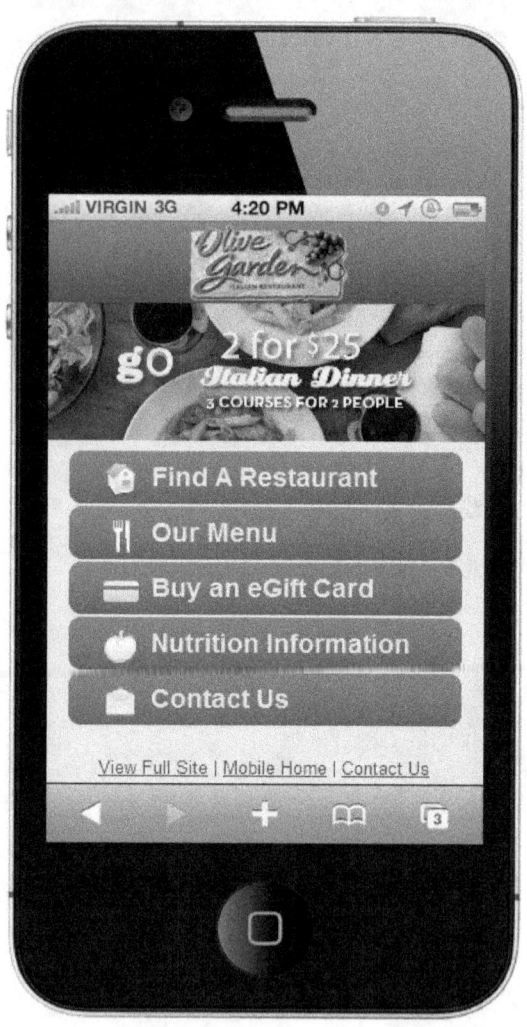

In other words, the website presented to your customer will be determined by the device they are using to browse the internet.

A mobile website is NOT meant to be a duplicate of every page, post and article on your regular website. If you'll notice, Olive Garden has responded to the top four reasons their customers look them up via mobile, plus an eGift card option for last minute gift giving during the recent holiday season:

- To find out where their nearest location is & get directions

- To see what's on the menu (and if it's compatible with the visitor's dietary needs)

- To look up nutritional information, perhaps for the purpose of avoiding allergens or helping a dieter stay successful

- Find contact information (e.g. phone number or email address)

- To buy a last minute eGift card

The web page is branded nicely with Olive Garden's colors and logo. There is a lovely header that mentions their newest offering – 3

Courses for 2 People for $25 — hard to pass up if you like their restaurant. And there are only five large "buttons" on the home page, leading visitors directly to the most commonly-sought-after information.

Everything is legible, clear, and easy to access and read, making this a great mobile site that people will actually come back to again as necessary.

We live in a fast-paced world where everybody is in a race with time and will not spend any extra time waiting for a certain website to load. So the aim of having a mobile friendly-website is allowing the many mobile phone users to access your business and get the information they are in search of — as quickly as possible.

What You Need on Your Mobile Site

To recap: No matter how web-savvy you are, your number one priority should be making sure your current website is mobile-friendly – either by using a responsive theme or design, for example a Wordpress theme or plug-in, or by creating a separate mobile site.

Thinking back to our Olive Garden model, here are the basic elements your mobile site should include:

- Your business's colors and logo

- Tap-to-call links or buttons

- 2-3 buttons leading to the most commonly sought information on your regular site or directly by telephone

- One or more social/viral elements

- One or more interactive elements

By *"social/viral elements,"* we mean anything that encourages the searcher or customer to pass your website links on to others.

This can include any combination of:

- "Forward to a Friend" link via e-mail, Twitter or Facebook

- Social sharing buttons for as many sites that you are active on

- Opt in registration page for a SMS or Text Campaign

An *interactive element* could be one of your existing buttons that goes to a coupon for a percentage off their next visit or a free appetizer with their next meal or a free car-freshener with their next car wash. It encourages participation on the part of your customer.

Meet Co-Author Brian Anderson

With over 20 years in software and internet technologies, Brian Anderson has been creating and delivering innovative digital marketing solutions to a diverse client base spanning both local and Fortune 1000 clients.

Brian has a B.S. from Florida State University and pursued his M.B.A at Notre Dame University. After a career in enterprise software, Brian turned his marketing passion and entrepreneurial spirit into building a renowned search agency, Peachtree SEO, focusing on local search for over 275 global clients.

Today, Brian spends his time running his mobile software and digital marketing companies as well as coaching and mentoring other marketing entrepreneurs. Brian is the President of Media Mash, a digital agency to the automotive industry, and is the co-founder and Managing Partner of MobileBizBox, a leading mobile marketing software platform for resellers.

SMS Text Messaging Marketing

"Text message marketing" refers to marketing strategies revolving around Short Message Services, commonly referred to as SMS. It is also a very powerful strategy known to link organizations and businesses directly with their customers around the clock.

When your kids spend half the day glued to their phones, thumbs flying "talking" to their friends, they are texting each other – yep, that's SMS.

What a lot of businesses aren't aware of is that the same type of messages can be used to market to their customers – sending them coupons, free incentives, appointment reminders, notice of an available time-slot with a busy stylist and more.

There are services available to businesses with features that allow them to send bulk messages to a list of recipients (your customer base). Using these services, it's as easy to send a text message to one person as it is to send one to 10,000 people.

SMS text campaigns are handled with keywords and shortcodes.

In the case of text messaging, "keyword" means the word you tell your customers to text *to* the shortcode.

The shortcode is the five or six digit number assigned to your business specifically for quick text messaging.

For example, if your customer reads the message "Text BURGER to 55555 to get a coupon," they would literally text the word "BURGER" (the keyword) – and only that specific word – to 55555 (your assigned shortcode). Almost immediately after the message is sent, your customer would receive their coupon for a free burger or discounted meal. So, since your customer (or potential customer) is right there – chances are they will come in and try out your menu.

Your shortcode would stay the same – it's your "address" – however, each new campaign/promotion or demographic you try would want different keywords so you can track your ROI easily.

The most common ROI estimate for text messaging success seems to be a solid three times that of other types of calls to action. One reason may be that people typically text at the point of sale – as buyers, they're already committed when they access your offer or coupon.

Text Messaging Cautions

There are some potential hazards you need to be aware of. You will be building a database of customers. And your customers do need to "opt in" (give you permission) before you can start sending them texts.

You also need to make sure you protect yourself (and alert your customers) by including necessary notifications in your text messaging ads and promotions. Anytime you cause your customer to incur a charge by receiving your text messages, you need to tell them that before they sign up and in the first message after they opted-in. Use a disclaimer such as "message and data rates may apply". If you are charging money for customers to text your keyword to your shortcode, you also need a clear notice letting them know precisely what amount they will be charged – and advising minors they are not allowed to participate, or not allowed to participate without parental consent, depending on what part of the world you are operating in.

You also don't want to overkill your customers with texts. Campaigns should be carefully spaced and scheduled, built around not only your target customer's "pace" but also around your sales funnel and significant events – either seasonal events or niche-specific ones. As a courtesy, you will also want to give your customers the maximum number of times you will be sending them a text message during a given month.

*"SMS (or text messaging) marketing is businesses communicating with customer via mobile phone with their **explicit***

permission. *And businesses need to do that at the right time, at the right place and while providing relevant value."*

The key here is that businesses are *communicating* with consumers – this is not just a one-way blast of SPAM. You, in return, will be providing something of value to them. This may be based on the knowledge you've acquired about your customers and that is the exciting aspect of this type of marketing. Customers have requested your text messages and that response will be received with anticipation, often resulting in action.

A Word of Caution: Mobile Marketing is a NO SPAM Zone! You do NOT want to send unwanted messages to anyone! If you think you can add your existing customer database into a text messaging campaign and then just start sending coupons, ads, etc. – think again. Don't do it! It's literally a $47 million dollar mistake to send your customers even one SMS message without their express consent – even if it's to give them something for free.

And here's a good case study as to why you never want to do that.

Read the full story at the link below, but what it boils down to is that [the parent company of] Jiffy Lube sent a coupon for 45% off an oil change to most customers who'd been their customer for a prior service. Jiffy Lube has a spot on their work order/invoice for a phone number and if it was given, then that customer got the text coupon. During the class action trial, the company tried saying that they had permission. However, the ruling came back that because the mobile number was given as a condition of the service being provided, it didn't count as consent for Jiffy Lube to market to them after the service was completed. **Ouch!**

http://www.tatango.com/blog/textmarks-sms-spam-lawsuit-what-it-means-for-sms-marketing/

Businesses have positively embraced the use of text messages in their marketing strategies and you should not be left behind. The reason why text message marketing is becoming popular is its affordability and ease of use. All you need is a list of opted-in customers and your message will be sent to multiple recipients with a push of a button.

SMS Features

Text messages have different features that makes marketing to your customers and opt-ins a completely different and effective strategy:

Bounce Ad

This is an automatic responder that will send a message when requested by a user. Businesses have to create pre-set messages that will be sent to mobile phone users when demanded. (Text "BURGER" to 83733)

Broadcast Feature

This is a service that enables businesses to send an unlimited amount of text messages to as many subscribers as possible. This can be easily done via a desktop or laptop.

Time Release Text

This feature allows business to create broadcast campaigns that will be used in future – think appointment reminders for services such as doctors, dentists, salons, spas, tire rotation, oil changes and more! The business then selects a date or dates when their messages will be sent to their subscriber(s). The feature allows businesses to save multiple campaigns therefore making work easier.

Appointment Reminders

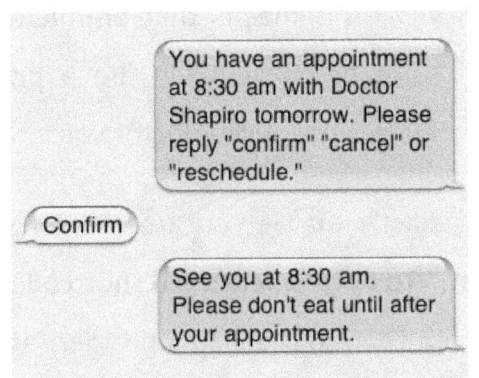

For those businesses who make their living by appointments, having an automatic reminder sent to customer via text message will free up an employee to do something other than leave voicemail and will likely cut down on missed revenue due to missed appointments.

Multi-Media Messaging (MMS)

This enables businesses to include customer images or coupons in text messages. This feature is effective especially when you want to promote a new item or when you want to attract more customers by offering coupons through text message marketing.

Dedicated and Shared

There are two types of shortcode: dedicated and shared.

"Shared" is when your customers text a number that is shared by many different businesses. This is usually the type of shortcode you'll be given if you sign up with a standard mobile advertising or text messaging service. The only real downside to sharing a shortcode is that all keywords are shared as well. This means that premium keywords such as "pizza" or "free" are probably taken. Be a bit creative and you will find plenty of keywords that will work for you.

A "dedicated" shortcode is exactly that – you are the only business that will have that "address." To have a dedicated shortcode you can expect to pay upwards of ten thousand dollars and there is usually a monthly fee to go along with that. For most small businesses, dedicated shortcodes are not a cost-effective strategy and there's no real benefit in having one except the vanity factor.

Your Open Rate

This refers literally to the number of "opens" compared to the number of text messages sent. Many Text Messaging providers confidently quote open rates between 95%-97% -- "5 times higher than email open rates."

Here's why. Text messages are:

- Often delivered at point-of-sale, when the customer is psychologically primed and ready to buy

- Immediate and real time

- Short

- Irresistible for most people – after all, it "only takes a second" to see what the text says

- Entertaining – whether people realize it or not. There's always that temptation to see if you're receiving garbage… or a nugget of gold

Text message marketing is simply the most affordable way to create product awareness and promote your small business.

Mobile Coupons Increase Foot Traffic

If you want to include mobile marketing in your marketing strategies, mobile coupon marketing is a great place to start.

A mobile coupon is not that different from the conventional coupons we are all used to. The only difference is that instead of being paper and something you have to cut out of a flyer or magazine, it is sent to customers on their mobile phones.

Give your customers a coupon for a discount or a free item/service to get them back to your store. Think about what would be an incentive that would get you motivated to drive to a store/restaurant/salon and spend your hard-earned money and then create something similar.

There are studies that show the potential of using mobile coupons, so this is a quick and easy way to venture into your new mobile marketing strategy.

To get ideas on the ways you can use coupons, type something like "mobile coupon" into your Google search bar and click "Images." You'll find a lot of coupon style examples to get you started on creating your own mobile coupons!

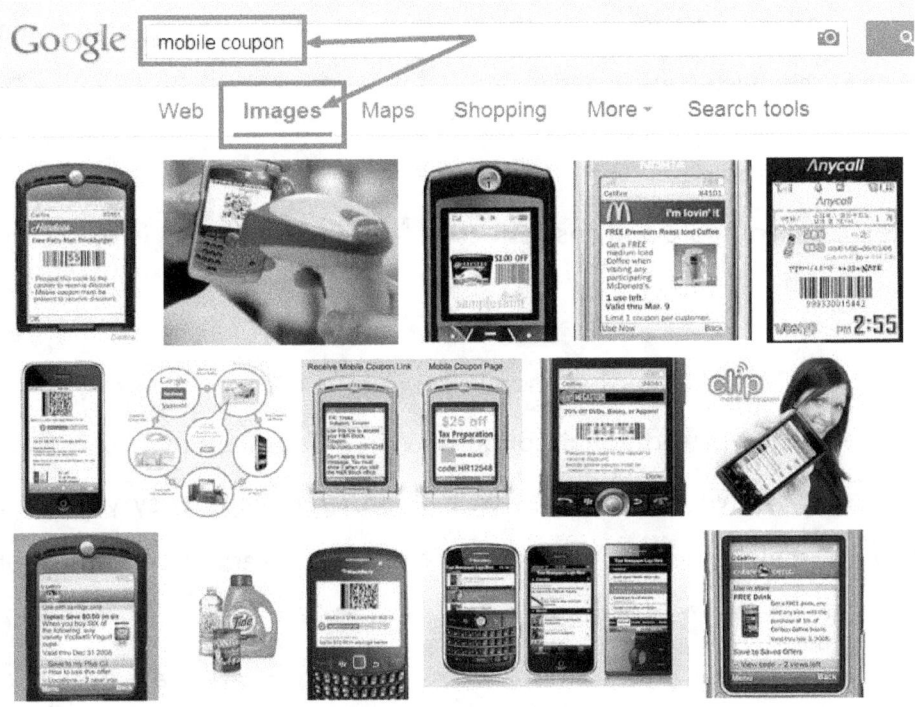

- Florists can use mobile coupons to remind their customers that Valentine's Day is just around the corner – and offer them a 20% repeat-customer discount.

- Automotive oil change shops can use mobile coupons to remind their customers when it's time to change their oil – along with a 15% off coupon.

- A restaurant that is having a slow night can offer a short-time offer – come in to the restaurant that same evening and receive a free appetizer.

Depending on the type of business, mobile coupons are one of the most profitable and cost-effective forms of marketing in existence today.

Statistics show that mobile coupons are **10x** more likely to be redeemed than traditional coupons. Why? Because your customers will always have their phones with them every time they visit your store. Mobile coupons are easier for your customers to keep track of and also easier to redeem than a traditional coupon. When the coupon is in their phone, there is less stress for everyone -- and your business will enjoy increased foot traffic.

More people are now interested in signing up for text message marketing and coupons are a huge plus to increase traffic to any particular business.

Another idea is to add social/viral element to the mobile coupon so that users can share your coupon with their friends and at the end of the day your site will have more visitors who will gradually transform to customers.

Meet Co-Author Susan Banks

Susan Banks is the founder and chief strategist at Mobile Marketing Toronto, and an unabashed lover of QR codes.

Susan's interest in technology began in 1998 after buying her first Windows PC. She started picking up computer magazines for fun (yes, sadly so) and decided to go back to school full time, taking a Network Computing course in 1999. Her first full-time job was at an Internet startup, start down. After weathering the first IT crash in 2000, she worked in various tech jobs until deciding to venture into self-employment and online sales and marketing in 2008.

In 2011, Susan realized that mobile was the place to be. Instead of using outdated, ineffective marketing methods, businesses could reach their customers where they were - at the end of their cell phones. And Mobile Marketing Toronto was born.

QR Codes *

QR codes are popping up everywhere; in magazines, on posters, on flyers that get delivered to your home, in store windows, and inside stores and restaurants. Whether or not your customers know what they are, chances are that they have seen a QR code before and will recognize one if they see it again.

To harness the power of QR codes in your marketing plan, it is important to know exactly what they are, what they can do, and how to use them effectively. This includes having an understanding of best practices as well as QR code fails. Since all QR codes are not created equal I will also cover some of the technical aspects of generating and tracking QR codes.

What Are QR Codes?

QR (quick response) codes are two dimensional (2-D) bar codes, similar to the UPC codes found on products in retail stores, but they can hold significantly more data than a regular bar code.

There are many different types of data that can be encoded in a QR code including a website URL, SMS text message, phone number, contact information and even just simple text. The beauty of a QR code is that it can be read by a QR code reader app on a smart phone. When scanned, the app will initiate the encoded action. For example, it will launch the browser and open the requested URL, pre-fill the text message, or load the phone number, so you just have to press a button to complete the action. QR codes bridge the physical and the digital world, and since anyone with a smart phone can scan a QR code, it is a very useful and versatile marketing tool.

QR codes are commonly seen in magazines and newspapers, and on print advertisements, television, websites, and billboards. They are scanned more in magazines than anywhere else. In an analysis of the top 100 magazines in the United States Nellymoser Inc. reported that "For the first time in Q2 2012, every magazine in our study printed at least one mobile action code. All but 10 of the magazine titles printed 10 or more codes in the quarter," and "The big news is that QR style mobile codes in magazines get higher response rates than other printed direct marketing tools." (Roger Matus, Executive Vice President, http://www.nellymoser.com/)

The number of American consumers scanning QR codes has increased dramatically in the last three years, from 1% in 2010, to 5% in 2011 (Forrester Research - http://www.forrester.com/) and 19% in 2012. In the 18-34-year-old age bracket this number is almost double.

http://www.pb.com/smb/qr-codes/marketing/statistics-trends-use/reports/2012-us-europe.

The main reason the majority of North Americans have still not scanned a QR code is lack of knowledge. They may not know what a QR code is. Or even if they have a smart phone and know what a QR code is, they may not know how to scan it. So, like any other technology, there is a learning curve involved in the adoption of QR codes. Education is the key.

However, once this hurdle is overcome, the ease and simplicity of scanning a QR code to access information, get a coupon, or enter a contest makes it the preferred choice of many mobile phone users.

Effective Use Of QR Codes:

QR codes make it easy to connect with your customers. They encourage user interaction and engagement, and can easily be integrated into your current marketing plan.

QR codes can be printed on a wide range of items including business cards, products, menus, store signs, billboards, T-shirts, key chains and other promotional products. They can be placed where it will be easy for the public to see or a potential customer to scan the code.

Before incorporating QR codes in your marketing strategy, make sure you know your objective. Who is your target market? What are you trying to achieve?

These are my top ten tips for the effective use of QR codes:

1. **Mobile optimized content:** If your QR code sends people to a web page make sure the page is mobile optimized. If someone is scanning your QR code they are on a smart phone. Do not send them to a non-mobile page. It will only frustrate and annoy them and make you look like an amateur. <u>This is by far the biggest mistake marketers make when using QR codes.</u>

2. **Call-to-action:** So many businesses attempt to show how current they are by printing a plain black QR code and sticking it in their store window, but with no information on where the QR code goes, what it does, or WHY anyone should scan it. This also applies to print ads where a QR code is stuck

somewhere on the page, almost as an afterthought. Give people a reason to scan your QR code and tell them what will happen when they do.

3. **Valuable content:** Reward the people who take the time to scan your QR code. This can be done by providing interesting relevant content, educating them, or offering them a coupon, discount, or even the chance to win a prize.

4. **Provide a text alternative** for those who prefer not to scan or don't know how. That could be a short URL to the web page, or the SMS text instructions. For example, a real estate sign could have one of the following:

**For Info & Virtual Tour
Scan QR Code
or go to goo.gl/aySsz**

**For Info & Virtual Tour
text 2kingst to 12345
or Scan QR Code**

5. **Provide Instructions**: Until QR codes become more mainstream it is worthwhile to educate consumers on how to scan them. For example: *"Need a QR code reader? We recommend i-nigma. Download from your app store or directly from i-nigma.mobi"* If you are providing a text alternative (as #4 above) it is not necessary to also provide instructions – but it doesn't hurt.

6. **Location:** Think about where your QR code is going to be scanned. Is there Internet access? What are the physical conditions? It must be bright enough to be able to read the QR code, but with minimal glare or distortion. Is the QR code large enough for the location (see "Tips for Creating Your QR Code" section below)?

7. **Take Advantage of a Captive Audience:** Utilize travel time. Buses, bus stops and train stations provide you with a captive audience. Give travellers something to do while they are standing or sitting around – often already looking at their smart phones. Subways stations and cars are not a good idea unless they have Internet connectivity. You might get a few people to scan your QR code, but they will not be able to access your site when they do. (Although virtually all QR codes

scanners will save the scan results, so you can access them when you are connected again, this tends to defeat the novelty of easy, instant access.) In South Korea, on the other hand, which does have coverage in its subways, retail giant Tesco Home Plus opened the world's first virtual store in the Seoul subway in August 2011. Each item in the store has a QR code, which when scanned will add the item to the purchaser's shopping cart.

8. **Survey Your Customers:** Find out what your customers think and want by asking them. A QR code can be used to direct your clients to an online poll or questionnaire; or for a more immediate response, poll them by a SMS text message.

9. **Contests:** People like to win stuff. In ScanLife's "Mobile Barcode Trend Report Q2 2012", ScanLife reported that contest campaigns generated the most scans.

10. **Tracking:** QR code scans can be tracked. You can find out how many people scanned your QR code, and even where they scanned it from. There are both free and paid services that allow you to create a unique tracking code for each QR code campaign.

Creative Uses of QR Codes:

The best QR codes are the ones that grab people's attention or generate curiosity. This could be because of their unique design, placement or what they are offering.

Well.ca Virtual Store:

There are lots of very creative uses of QR codes. One that grabbed my attention was a QR code virtual store located in Toronto's PATH.

"PATH is downtown Toronto's underground walkway linking 28 kilometres of shopping, services and entertainment.....More than 50 buildings/office towers ...Twenty parking garages, five subway stations, two major department stores, six major hotels, and a railway terminal are ... accessible through PATH"

(http://www.toronto.ca/path/)

Living in Toronto, I am quite familiar with PATH, and regularly navigate this underground world, as do **more than 100,000 daily commuters.**

In April 2012, Well.ca (http://well.ca/), Canada's largest online health and beauty store, opened a virtual store in Brookfield Place in the PATH.

Pedestrians could order products by downloading Well.ca's app and scanning the QR codes beside the actual products. Their orders would then be delivered to their homes, often the next day.

Well.ca took advantage of a large, captive audience. Since the store was located in the financial district, a large percentage of the people passing their virtual store owned smart phones. The store was beautifully laid out and all the products for sale were on display with the QR code beside them. The shopping experience was unique and simple.

The virtual store was a huge success. "All thirteen brands that appeared in the Virtual Store experienced at least a 100% increase in sales during the campaign. One brand experienced a 1,000% increase over the previous month, selling more during the campaign's first week than in the previous 12 months combined. The products featured in the virtual store continued to grow post-campaign (26% over pre-virtual store time period), demonstrating the sustained value of the customers acquired during the campaign." (Source: Well.ca)

When asked if they would be opening another virtual store, Senior VP Paige Malling said "We're always exploring new ways to impact the marketplace, so more Virtual Stores are never out of the question!"

QR Code Corn Maze:

The Kraay Family Farm near Lacombe, Alberta, Canada has over 20 attractions for families including a 15-acre corn maze, a farm yard, mini golf, rides and games. Over the past 13 years the Kraay family has been creating original designs for their corn maze, including logos for national sports teams.

In 2012 Rachel Kraay "was just relaxing, reading a magazine and saw a whole bunch of QR codes and I thought, you know, it looked a whole lot like a maze I wonder if we can make one," (CTV News). So, with the help of their designer they did.

Their QR code corn maze, which worked when scanned from the sky, sent people to the Kraay Family Farm website (http://www.kraayfamilyfarm.com). The Kraays submitted the seven acre maze to Guinness World Records, and on July 24, 2012, Guinness certified the 309,570 square foot (28,760 square metre) QR code corn maze as the world's largest QR code.

Although not many people would be flying overhead to scan this particular QR code, the idea was unique, generated a lot of publicity, and was recognized by Guinness. So even when the Kraay's QR code maze is replaced with another design, the pictures, the story, and world record remain.

Victoria's Secret:

In 2011 Victoria's Secret had a very seductive billboard ad campaign where they strategically placed white rectangles with QR codes on photos of their lingerie-clad models so that only their skin was visible. Below the picture of the model were the words "Reveal Candice's (or Erin's or Lily's) Secret." To see what was behind the QR code you had to scan it. You were then taken to the same picture of the lingerie clad model, less the white bar. This "Sexier than Skin" campaign generated a lot of curiosity and interest.

Scandinavian Airlines:

In January 2012, Scandinavian Airlines ran a "Couple Up To Buckle Up" 2-for-1 promo. It sent emails to 100,000 of its customers with two QR codes, encouraging them to book a flight with their loved one. Each QR code went to a separate YouTube video. Only by watching both videos at the same time and placing the videos side by side could they get the coupon code for the 2-for-1 flights. The flights sold out.

myToys.de:

The German toy store, myToys.de (http://www.mytoys.de/) used Lego to build QR codes, which when scanned sent customers to their online store. The different, colorful QR codes, drove users to pages on myToys.de's website where they could purchase the Lego set used to build that specific QR code. While the campaign was live, 49% of visitors to the myToys.de website came via the QR codes and twice as many brick boxes were sold for the Lego models included in the QR adverts.

Taco Bell:

Taco Bell ran a very successful QR code campaign in December 2012 aimed at sports fans. Taco Bell teamed up with ESPN sports, and put QR codes on their taco boxes and soda cups. The QR codes sent users to a video of curated content from ESPN with pre-game analysis for the Bowl Championship Series college football series. The QR codes were large, in the middle of the packaging, and instructed their customers to "Scan The Code" in large letters. Over 225,000 people scanned these QR codes during the campaign which ran from December 20, 2012, to February 3, 2013.

QR Codes – Not All Paper And Ink:

QR codes in unexpected places generate curiosity.

After much trial and error, Moshi Moshi, a sushi restaurant in England has perfected a sushi wrapper with a QR code. The QR code, made out of rice paper and squid ink, will take their customers to website where they find out where the sustainably farmed fish they are eating has come from. (http://www.moshimoshi.co.uk/)

In Rio De Janeiro officials have started embedding QR codes into pavements to guide tourists around the city. QR codes are also being engraved on tombstones, dog tags, and on human bodies as tattoos.

Guinness beer ran a campaign in 2012 where they put QR codes on pint glasses, with could only be scanned when the glass was full of dark beer - preferably Guinness!

Have fun and be resourceful when creating QR codes. Generate curiosity and you will be rewarded.

QR Code Fails:

When planning a QR code campaign it is just as important (maybe more) to know what doesn't work as what does. If you just Google "QR code fails" you will be provided with an endless supply of bad ideas, and some entertainment to boot. A good place to start is http://wtfqrcodes.com/.

It is not hard to find very poor implementation of QR codes, both personally and on online. Here are some examples. To protect the guilty, images and company names have been withheld (by me).

QR Codes On Subway Cars

To find lots of bad examples of QR code usage I just have to get on a subway car.

One ad for a local community college in Toronto – one that ironically teaches technology courses – had all the following problems:

- Ad was above the seats, on a curved surface, which was not well lit.

- QR code was small, about 2" square, which under ideal conditions could be scanned a maximum of 20" away. Because of the placement and lighting it could in fact only be scanned from about less half that distance.

- In order to scan the QR code I had to lean over the people in their seats, hold my phone seven feet in the air and almost a foot from the actual QR code. Who would ever do that?

- Since we were underground with no Internet access, I could not even check the website until I left the subway station. When I did it was not mobile optimized. This was in fact the case with most of the QR codes I scanned on the subway.

What were they thinking??

QR Codes In Public Washrooms

Ok, now I know you have a captive audience here, but don't you think they have other things on their mind? I have seen photos of QR codes in men's urinals, and personally have seen them inside the stalls of women's washrooms and on the hand dryer beside the sink. Yes, I am glad your hand dryer is so green, but no I am not going to pull out my phone and scan that tiny QR code on it to learn more.

Survey In Grocery Store - Nice Try, But ...

As I entered my local grocery store, there was a large poster asking "How Did We Do" with an option to scan a QR code or visit a website to complete a survey and have the chance of winning a $200 gift certificate. It was clear on the poster what was being requested, there was incentive to scan the QR code, and the QR code was large and scanned easily. So, what was the problem?

After scanning the QR code, selecting my language and accepting the conditions the progress bar on the top showed I was 1% complete. I answered the first question - 2% complete. Second question - 3% complete. Seriously? How many questions are in this survey? I did not have the patience to find out. I am wondering how many people scanning the QR code would have even got this far. Think of your audience! Do you really think people grocery shopping are going to stand there and complete a survey that tells them after four steps tells them they are 3% complete?

QR codes On Retail Products

I think QR codes on product packaging is a great idea - but please - send your customers to a relevant mobile site. I bought a new USB headset for my computer and on the packaging was a large, easily scannable QR code. I thought I would be directed to a mobile page with information on my new headset, but instead was directed to the manufacturer's main, non-mobile site. Why?

Rule of Thumb:

Before using a QR code in your marketing think about whether it makes sense and whether you, as a customer, would actually scan it. If you are in doubt, test it on some real people, especially if you are going to be printing them out. Better to revise your strategy than having thousands of ineffective QR codes out there in the wild.

QR Code Technical Basics:

Although it is not necessary to know all the technical details about how the data is encoded in a QR code, there is some basic information that will help you when generating and using QR codes.

Density:

QR codes contain encoded data. In fact they can contain a lot of encoded data – 200x more than standard bar codes. The more data encoded, the denser the QR code. The denser the QR code, the more difficult it is to scan.

In other words, if you are going to be using QR codes in your marketing material you need to be aware that there is a trade-off between the amount of data that can be encoded and the scannability of the QR code.

Error Correction Code (ECC):

QR codes use a sophisticated error correction algorithm which allows QR codes to be read even if they are dirty, damaged or distorted.

QR codes can be generated at four different error correction levels: L (7%), M (15%) Q (25%) and H (30%). The lower the error correction level the less data is encoded in the QR code, and the greater distance it can be scanned. Higher error correction levels increase the scannability of the QR code under less than ideal conditions. The percentage refers to how much of the QR code can be damaged or distorted while still having a scannable QR code.

There is no hard and fast rule for which ECC to use, but if you are not sure, M (15%) is a safe bet, unless you are creating a custom QR code, in which case you should always use H (30%). For small QR codes, for example on a business card, a L (7%) ECC is recommended.

QR Code Generators:

To create a QR code, you need to use a QR code generator. If you search the Internet there is no shortage of QR code generators. Are they all created equal? The short answer is no.

Everyone has their favourite QR code generators, and like other tools, it is not always one size fits all. Here are a few of my favourites.

1. <u>GoQR.me</u> is the one I use most frequently, by far. Why? Because it is clean, simple to use and the URL is easy to remember! Yes – that matters. Also, it has lots of options, including changing the error correction code (ECC), the color of the QR code and the background, and the size of the margin around the QR code.

   It is limited in the type of QR code you generate to: Text, URL, Call, SMS and vCard, but since this covers the majority of QR codes generated, I don't really see it as a drawback.

2. <u>QRickit.com</u> is another great free QR code generator that offers a lot functionality. You can create many different types of QR codes with the following options:

   - Change color of QR code, background and text;

   - Generate .jpeg, .png or .gif images in various sizes;

   - Create high resolution 2400x2400 pixel .png and 300 dpi .jpeg image files for graphic design and high-quality printing.

3. <u>QRstuff.com</u> is another popular QR code generator. It can generate 23 different data types of QR codes, and has both

free and paid versions. The free version only allows low ECC QR codes (7%). Paid allows generation of large print quality vector image QR codes.

Tips For Creating Your QR Code:

- **Short URLs**: If the QR code you are creating is small and the URL you are sending people to is long, consider using a URL shortener before generating the QR code. This will minimize the data and increase scannability.

- **Margin:** Make sure your QR code has a white, or light color margin.

- **Size**: Is your QR code larger enough to be scanned from the distance people will actually be scanning from? Under ideal conditions a QR code can be scanned a maximum of 7- 10x the width of the QR code. So a 6" square QR code can be scanned from a maximum of 42-60". Although it is possible to get away with a smaller QR code if the URL is short, as a rule the minimum size of a QR code is 1" square.

- **Context:** On what material will the QR code be printed? Where will it be scanned? What will be the conditions under which it will be scanned? Under poor conditions make sure you use an ECC of Q (25%) or H (30%).

- **Test:** Not all QR codes will scan, so test, test and test again before you print, ideally using more than one QR code scanner. Also, test under poor lighting conditions, and at the actual distance people will be standing when scanning the QR code.

QR Code Readers

In order to scan a QR code with a smart phone you need a QR code reader app installed. Although some smart phones have readers installed, there are many excellent QR code reader apps that can be downloaded.

My QR code reader of choice is **i-nigma**. It works on all the major smart phones on the market, is quick and very responsive. It can be downloaded directly to your smart phone from

http://i-nigma.mobi or from your phone's app store.

Beyond Black and White

Standard QR codes are black and white. They offer optimal contrast for fast scanning. The problem with black and white QR codes are that they are boring and are not branded.

Custom QR codes provide businesses another avenue to brand themselves. You can embed an image or logo, and match your company's colors. They also add a level of comfort and security for the user, especially if the QR code is from a business they know and trust. In a poll by http://QRlicious.com, 88% of the people surveyed said they would be more likely to scan a custom QR code than a black and white one. They are more engaging – which is what you want.

When creating a custom QR code, it is necessary to generate the base QR code at ECC (error correction code) H or 30%, to ensure it will still scan after the customization, which may include distorting the basic design and embedding a logo or other graphic in, and/or behind the QR code. And don't forget to test to make sure the QR code still scans. Preferably on different QR code readers, and if possible on different types of smart phones.

Still, custom QR codes are not always the best. If the QR code is quite small, there is a lot of data encoded, or the scanning conditions are poor (eg: poor lighting, glare, uneven surface, outdoors,) a black and white QR code is a better bet, since it is easier to scan.

If customizing a QR code, try to minimize the data encoded. No point having a beautiful QR code that cannot be scanned.

QR Code Tracking:

When using QR codes in your marketing arsenal, it is important to track how many times they are scanned. This will allow you to determine your return on investment and fine-tune your campaigns.

There are several ways to track QR codes, including paid services and URL shorteners like bit.ly and goo.gl.

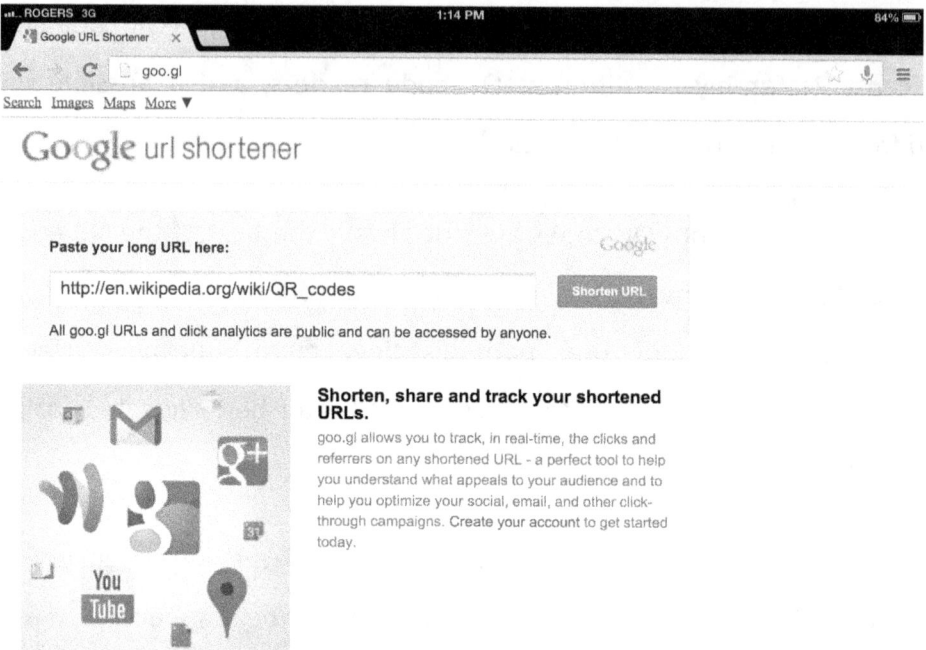

My favourite free URL tracking tool is goo.gl, because it does all the following:

- Shortens URL

- Creates QR codes

- Tracks QR code scans, providing detailed analytics (it actually tracks the number of clicks on your link – but you can create a unique link for each QR code.)

Also, as per Google, goo.gl short URLs are stable, secure (warning message if the short URL points to a suspected malware, phishing, or spam website) and fast.

And because it is Google, you can be pretty confident that they are not going out of business any time soon!

If you have a Gmail (or Google Docs hosted email) account, you can login to create and track QR codes using goo.gl. If you don't have one already you can create a free Gmail account. After logging in, go to goo.gl, paste in your long URL and press the "Shorten URL" button.

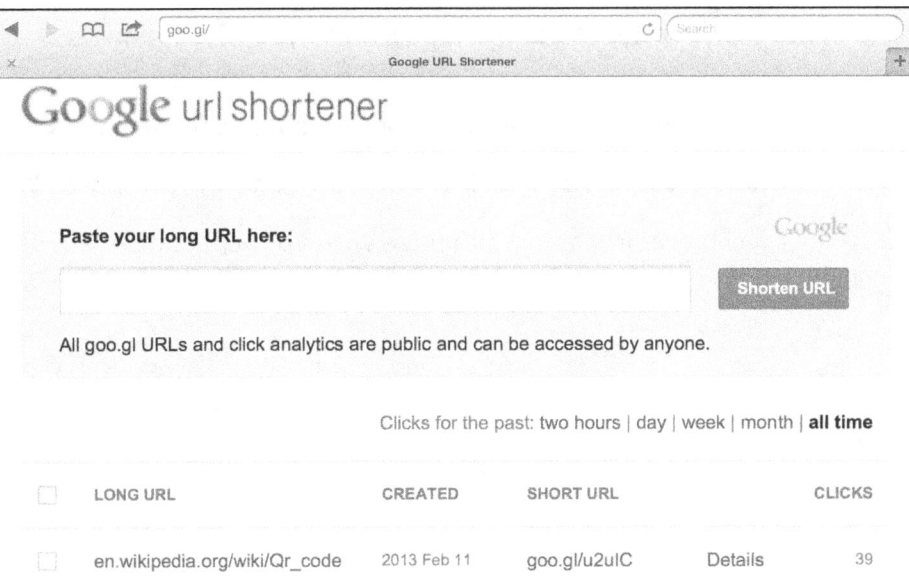

Create QR Code:

After creating a shortened URL using Goo.gl, there are two ways you can access your QR code.

1. **Add .qr to the end of the short URL.** Copy the short URL into the address bar and add .qr at the end and press enter. For example, goo.gl/u2ulC.qr

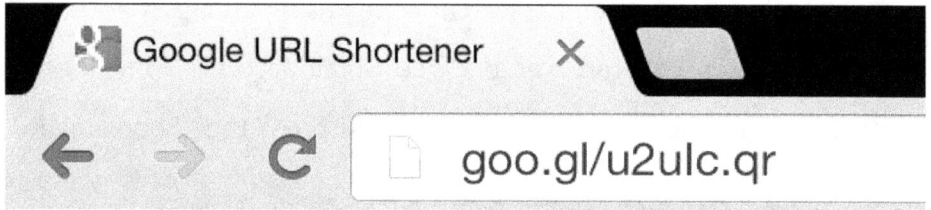

You will be redirected to the page with your QR code.

2. **Click the "Details" link beside the short URL on the goo.gl home page.** Then click on the QR code on the page. You will be redirected to the page with your QR code.

chart.googleapis.com/chart?cht=qr&chs=150x150&choe=UTF-8&chld=H&ch

chart 150×150 pixels

3. **Right click on the image to "Save image as", and save it to your computer**. This will save it as a 150x150 pixel .png image. If you want a different size, just change the numbers (150x150) in the address bar to the size you want (to a maximum of 500x500). In the example below I have created a 250x250 pixel QR code.

chart.googleapis.com/chart?cht=qr&chs=250x250&choe=UTF-8&chld=H&ch

chart 250×250 pixels

Analytics and Tracking:

All short URLs created while you are logged in to your Gmail or Google Docs account can be tracked. If using the short URL to generate the QR code, you can also use this to track the number of times your QR code has been scanned, where, and on what devices.

After logging in, go to the goo.gl home page. All your short URLs will be listed, sorted in chronological order.

Click the "Details" link beside the short URL or copy the short URL into the address bar and add ".info" at the end and press enter. For example:

http://goo.gl/u2uIC.info will redirect to the analytics page for the short URL.

One disadvantage with using the goo.gl URL shortener is that the analytics is publicly available to all users. However, the information of the user who generated the URL is not public.

Although you can shorten a URL and create a QR code without logging into your Gmail account, the shortened URL generated will not be specific to you. In this case, the same short URL is reused each time a long URL is shortened, across multiple users.

For example:

If not signed in to Gmail, the short URL generated for http://en.wikipedia.org/wiki/QR_code will be goo.gl/op1H. The tracking goo.gl/op1H.info will be for all non-signed in users.

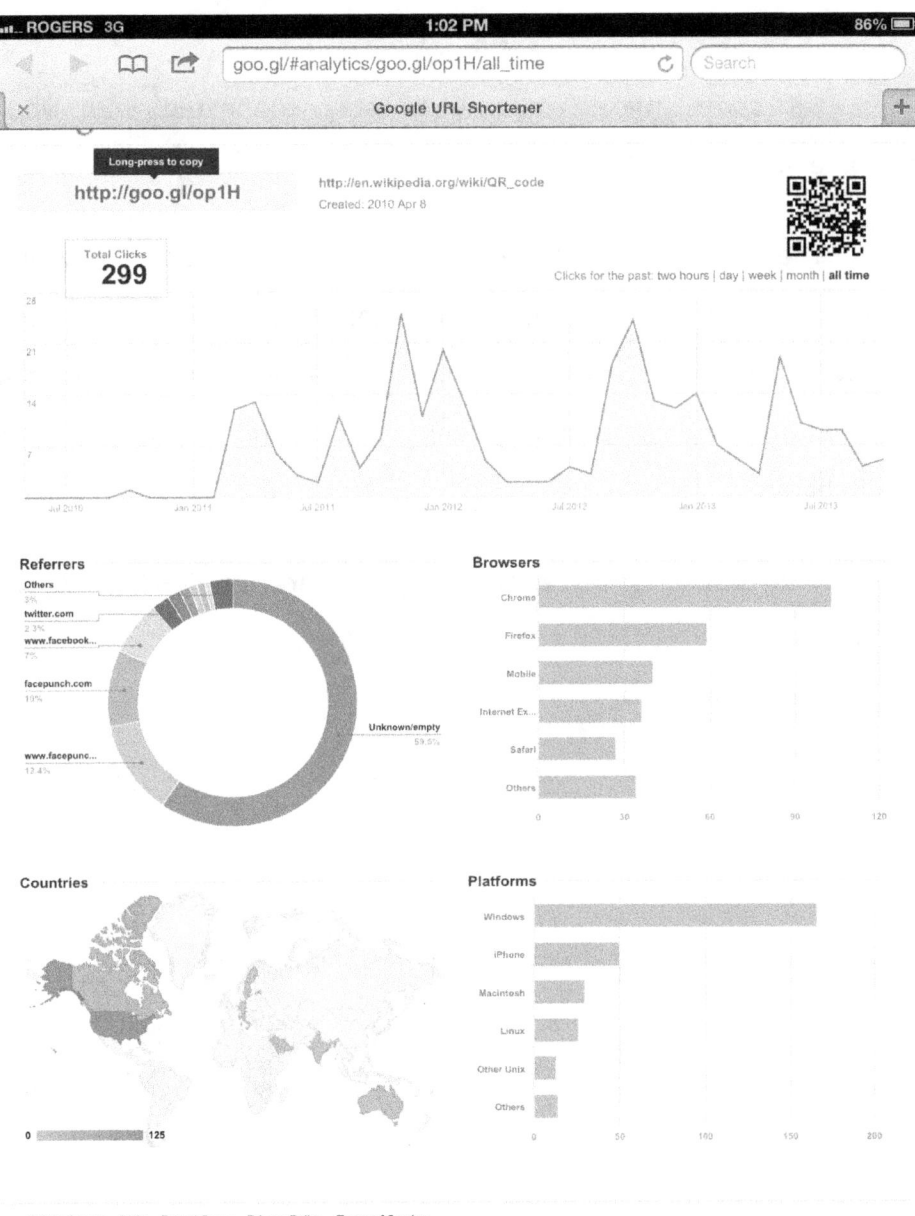

Final Words About QR Codes:

QR codes makes it easy to stay connected with your customers.

The simple black and white 2d bar code takes on new life when used in your marketing campaigns. With a little color and creativity, there is no end to the number of ways that QR codes can be used to increase customer engagement and ultimately, your profitability.

*QR Code is a registered trademark of DENSO WAVE INCORPORATED

Meet Co-Author Jeffery Cepuran

Jeffery Cepuran, a three time Amazon Best Selling Author, was born in Orlando, Florida on November 14th. Jeff was the second oldest of five children. Jeff attended college at Seminole Junior College and Florida Technological University. While in college he served as a university facilitator helping new college students acclimate to the college life and tutoring those having learning problems. He was a member of the Delta Tau Delta and Beta Delta Alpha fraternities and Alpha Phi Omega national service fraternity, the rowing team and the Air Force ROTC program.

He graduated from Florida Technological University (University of Central Florida) with a B.S. in General Studies and was also sworn in as a Second Lieutenant in the United States Air Force. There was a delay on his entry to the military, so Jeff continued his college education getting a B.A. in Secondary Education, with minors in math, physical science, physics, radio and television, and aerospace studies the following year.

Jeff served in the United States Air Force for 12 ½ years as a maintenance officer and a pilot. While in stationed in Omaha, NE, Jeff met and married his wife Marian, September 17th. After Jeff was discharged from the military he started working as an airline pilot at Comair Airlines, based at the Cincinnati Northern Kentucky Airport in 1988. While Jeff was in the military he became interested in the internet after purchasing an IBM PC with an 8088 processor. He started doing some marketing and sales honing his skills while still in the military.

Jeff started becoming actively involved in online marketing in 2010. After a year Jeff still hadn't made any revenue from his online businesses. Jeff has always loved meeting and talking to people, and after talking to some offline marketers, Jeff began his internet consulting business in 2011. On September 29, 2012 Jeff's airline was closed down and Jeff has become a full time offline internet consultant. Jeff's business continues to grow and in December 2012, he decided to incorporate and JC Internet Solutions, Inc was formed, *"Our Business is to help Your Business with Your Internet Solutions"*.

For the past year, Jeff has authored six books four of which were best sellers and one ranked as the "Book of the Month". He has been asked to beta test several internet products, be a mentor for many new offline business associates, while continuing to build his offline business.

Mobile Marketing and Social Media

Whether you personally indulge in social media networking or not; **YOUR CUSTOMERS DO.**

Mobile marketing works perfectly with social media sites such as Twitter, Facebook and FourSquare.

Social trends such as "check ins" allow mobile sharing to take on new significance. A "check in" is when a customer posts on their social media page that they are at your store or business – some social networking platforms have apps that can be installed just for that purpose such as http://Foursquare.com or http://www.Yelp.com:

Yelp describes itself as "the fun and easy way to find and talk about great (and not so great) local businesses."

And someone telling the world they're visiting your store, restaurant or company on a social network can be as good as a recommendation!

It might be a good idea to add a couple of questions to your customer questionnaire/survey about their social networking preferences. At the same time, you will actually be educating them by bringing platforms such as Yelp and FourSquare to their attention.

Restaurant Case Study

Restaurants have been the leaders when it comes to embracing Mobile Marketing.

Mobile promotions are ideally suited to restaurant needs – it's easy to add advertising tents to the tables and present an irresistible offer of a free dessert or appetizer when they text to sign up for a VIP list.

It's instant gratification when the item is offered for the meal the customer is currently enjoying. In addition, being able to send "last minute" texts on slow nights help fill an empty restaurant. Imagine being able to contact your entire customer base to let them know that "kids eat free tonight only" or "buy one get one free dinner – tonight only."

We've already looked at the example of The Olive Garden restaurant chain, so let's look at one that The Mobile Marketer recently presented as a video on YouTube...

Pizza Hut created a contest, with the objective of making their pizza and pasta dishes stand out from other heavy competitors with a similar product line.

They presented a four-step approach:

1. Pizza Hut created a contest, based on the keyword "WINPIZZA" (please note that this spelling is not the actual keyword used)

2. They loaded the contest with tempting prizes – not just a free slice of pizza, but prizes like "free Chicken Alfredo for a year."

3. They also made use of the "Forward to Friend" mobile option, allowing entrants to forward to five friends for an instant prize of a 2-liter Pepsi

4. They created a TV commercial that quickly let people know about the contest, and how to enter ("Text WINPIZZA to 55555")

They chose this approach because they were targeting a specific demographic: both male & female, ages 18-35 – due to the research that this demographic was the one that never left home without their phones and were most likely to own a smartphone. These were the customers that Pizza Hut wanted to add to their database.

The results? 3,000 people entered the contest during the commercial alone.

At the end of the promotion, Pizza Hut had a mobile database of 12,000 hungry customers. The results were much higher than expected, which they attributed to the "Forward to a Friend" option. In addition, Pizza Hut added a great tag line for their contest: "Keep your eye on your phone" – an irresistible command to mobile users addicted to their phones.

A final word on Pizza Hut: They now have mobile apps available for Android, iPhone, iPad & Windows phones to make choosing and ordering even easier for their enthusiastic mobile customers.

(Notice the QR code added for good measure underneath the App icons)

Don't underestimate the power of mobile marketing, when it comes to promoting your restaurant. If you scan various mobile promotion companies, they all have their own variation of projected ROIs similar to this one:

If you have:		And you send:		And you get a:		And they spend:		You will make:
500	X	4	X	15	X	20	=	$6000
opt-in customers		messages per month		*percent redemption		dollars per offer		dollars per month!

Restaurant Mobile Strategies

These strategies have proven to work well for countless other Pizza restaurants.

SMS Text Messaging

This allows people to text in orders before they arrive at the restaurant, perfect for people who are on their evening commute back home. Why not add a coupon to that mix -- many pizza chains successfully have. You can also target your texts to specific times of day – for example, sending one out before your peak meal times (or sending out an "early bird special" text to attract customers in during your slow periods just before a meal time rush).

Mobile App

Providing an app that allows customers to pick specific ingredients is another tool pizza chains in particular have used with great success.

QR Code "Weekly Special"

Another way to promote your restaurant and build customer loyalty: Use a QR code on a physical flyer, fridge magnet or other promotional item. Your customer scans the QR code to quickly see what the weekly special is – and order it for pick-up.

Although most people still prefer to text QR codes have the advantage of capturing hard-to-get customers by allowing them to be "clever" in front of their friends. (Scanning a QR code, if they've installed a QR code scanner app and are educated about how they work, seems to be **treated as a way to impress your peers**. It's interactive – you do something clever and get a reward. And the interactive reward factor is often the trigger for picky customers.

QR codes can be placed on just about anything that is given to your customer, placed on table tops or stand by the register: posters, coasters, flyers, fridge magnets and, of course, printed on the pizza box. Don't forget using them in social network ads – which have the advantage of being shared virally. A happy customer loves to share great restaurants they find!

Real Estate Case Study

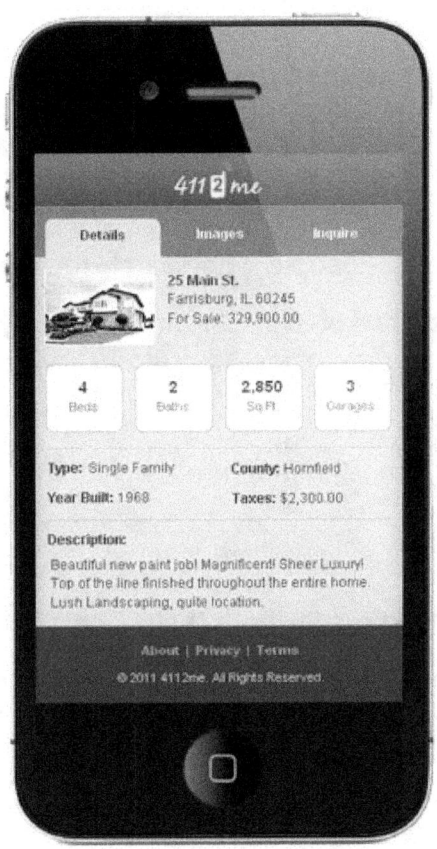

Real Estate is another business where mobile marketing is the perfect fit.

Have you ever been house-hunting and find a gorgeous home with one of those boxes where real estate agents keep flyers with information about the house. How many bedrooms, square footage, if it has a pool and when the kitchen was last updated.

You know the ones I mean.

How many times have you actually opened the box and found flyers inside??

Myself? ……. **Never!**

The @#$* boxes are always empty and it's infuriating!

We all want instant gratification and that includes instant listings while out scouting for likely properties.

More and more home buyers like to feel that they are in control of the buying experience – not the real estate agent. Not to mention that there's a little-recognized fact that's being noted by a growing number of real estate agents -- people who search for real estate via mobile tend to be in the bracket that can afford to buy, their time is precious and they mean business.

Here's a short study of a campaign mounted by Coldwell Banker in Pittsburgh, PA.

Coldwell Banker wanted to save time for both their agents and buyers. So their goal was to streamline the buying process and give individuals selling their homes more options. Their target market was ambitious: potential buyers, sellers and renters.

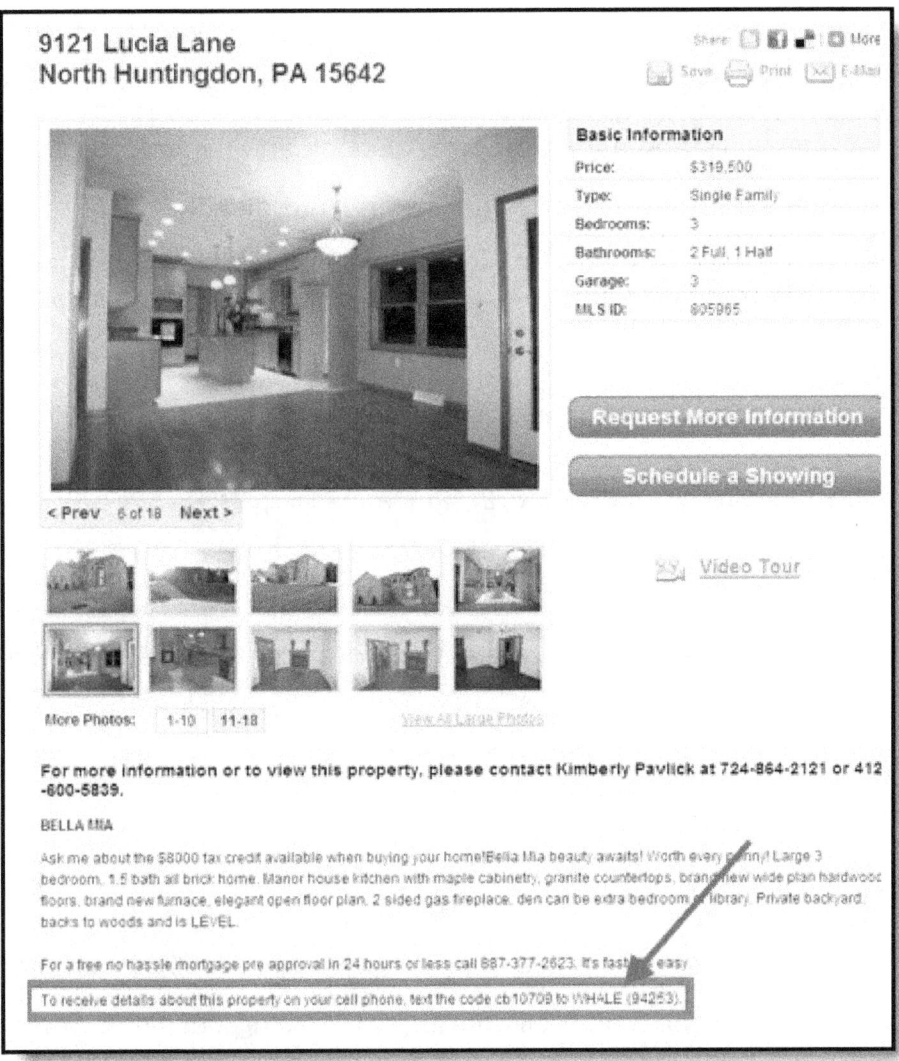

Their three-step approach used SMS text messaging.

1. They invited consumers to text the keyword "WHALE" to their shortcode number. This opted the customers in to receive text messages about this property. Customers instantly received the auto response text with property information.

2. They provided Calls-to-Action and alerted people to this option by means of newspaper ads, ads on their website and by inserting a sign rider and specific house code for each house

3. They tested this campaign out on over half their listings in Pittsburgh, PA.

The results? 94,186 visits to their Home page and 1,816 texts for the sign riders.

Real Estate Mobile Strategies

Text Messaging

You can combine these in different ways, as well as use texts to remind clients or sellers of showings. Allowing people to text a shortcode and quickly see if a listing is right for them can be a highly effective time saver in that it pre-qualifies potential customers – and brings in the right ones.

Add Texting & QR Calls-to-Action

A quick and inexpensive way to have the properties you're listing found via mobile is to run texting campaigns and use QR codes for your listings. Consider this when your listings are not getting found or are getting drowned out by other listings.

Virtual Business Cards

These seem to be particularly effective for real estate agents – especially when you include a QR code to all your listings. This will help potential buyers pre-qualify themselves so that buyers not in your ballpark don't contact you. It also shows your potential clients that you value their time and you are proactive. Your virtual business card can send people searching by mobile device directly to your mobile website.

QR Codes

Many real estate professionals have already discovered the value of adding QR codes to traditional signs and flyers, business cards and social media ads such as those found on Facebook.

Crucial Tips

- Both mobile phones and the ability to send and receive texts are especially crucial for real estate agents. According to the March 2011 business review, *real estate agents that responded to inquiries within one hour were seven times more likely to capture the lead* than those that took more than an hour to follow up and respond.

- If your normal client demographic tends to be young professionals under 40, with a college education and income over 100k, mobile marketing is particularly essential for you.

- If your normal client demographic tends to be over 45 with families, keep it simple. Statistics show they are more likely to telephone from a land-line or use personal computers.

Salon Case Study

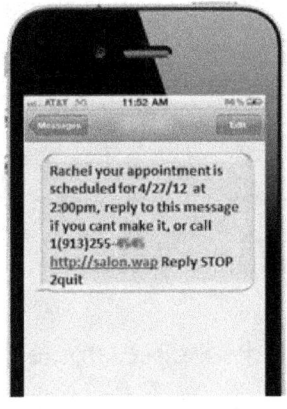

Hair salons are another ideal match with mobile marketing. We'll use a second study case presented widely across the net: that of Headrush, a United Kingdom hair salon.

Their goal was to reduce the number of no-shows – a common problem for many hair salons. Not only do no-shows cause lost revenue for the salon (and the stylist-depending on how they are paid), it can also lead to poor morale and professional tension or jealousy.

Headrush used a simple two-step approach:

1. They let their customers know they could receive appointment reminders by text message.

2. The day before each appointment, the customer was sent a simple text reminder, which acted as a call-to-action, getting these customers focused on and aware of their appointments again

Headrush exceeded its objective, reducing the number of no-shows by 70%. Their existing clientele consisted of 4,700 customers.

None of the reports detail how they made their customers aware of their text reminder program, but any salon could expect success by adding details clearly to counter displays, emails, flyers, postcards, direct letters and radio or TV ads.

Mobile Apps

Mobile Applications (better known as "Mobile Apps") have been causing a major buzz in the mobile device world for a couple of years now. Mobile software developers are continuously creating software aimed at making the lives of mobile device users more interesting, entertaining and convenient. Mobile device users encourage that by constantly searching for mobile apps that are fun, entertaining and helpful.

Mobile apps instantly connect mobile users to Internet services that are normally accessible by desktop computer. These apps give smartphone users a way to carry out different tasks on their mobile devices faster, easier and without a lot of hassle.

Some examples of mobile apps would be for social media sites such as Facebook, LinkedIn, FourSquare, and Twitter. Also, access to services such as Pinterest & Instagram for photos; Kindle apps for downloading and reading books – and lest we forget, Angry Birds for when we are bored and want to play a game.

The first mobile app was created by Apple for their users but with time, software developers ventured into the most popular smartphones operating systems; making apps easily accessible to the whole world.

Most people use mobile apps without even knowing it. For instance, if you have a smartphone and you use a mobile phone browser, then you are using a mobile app. Some mobile apps are pre-installed on smartphones, such as access to a "Mall" where you can download and purchase more Apps, calculators, e-mail, calendars, games and more.

People are addicted to mobile phones and apps play a major role in this. For this reason, businesses are quickly changing their marketing strategies and creating space for mobile apps in their marketing campaigns. There are mobile apps that have already made a name for themselves, such as social media sites. But this does not mean that you cannot have one created for your business and get the same benefits.

With the number of mobile device users increasing and communication companies designing smarter phones every day, one of the best ways to reach out to people is through mobile apps. Mobile apps are an important part of the future for any business.

Therefore, it is crucial that you evaluate your business to see how mobile apps can help you increase customers, sales, and profits.

How Mobile Apps Can Increase Sales

The most important part of any business is effective marketing. Business owners are determined to create brand awareness and let as many people as possible know that his or her business exists.

With the availability of affordable smartphones and millions of amazing mobile applications, everyone literally has their eyes glued to these gadgets. For business owners, this is a great opportunity to effectively and affordably market their businesses. If you want to increase sales, you should know what mobile apps are and their benefits to your business.

The main aim of having mobile apps is ease and simplicity. Give this to your customers, and watch your business flourish. If your customers realize that they can easily find you without having to go through countless hoops on their mobile phones, your business will always be on their radar.

A mobile app allows your business to be available right at their fingertips, which will bring your business closer to your customers. With just a simple click, they can get what they want from you. When your customers can give you a call or visit your website with a quick press of a button, they will become happy campers.

Mobile apps allow your customers to instantly access your business via their mobile devices without much hassle and this will work positively towards increasing your sales. People like being connected to businesses and if you can provide that much-needed connection, you will be rewarded with loyalty from your customers.

Mobile apps also include push notification which is very effective. Push notification is basically a text message that is sent to your customers via the mobile app that they have installed on their mobile device. Only about 20% of the emails you send your customers will be read while 98% of text messages are read within seconds. This means that by using push notification you can grow your business by generating repeat sales over a short period of time.

It's always a good idea to include coupons and promotions in your mobile app. Chances are that anyone who downloads your app will visit your location for the discount or freebie. They will have a chance to see what else you have for them and, who knows, might buy something else.

If you have a lot of interesting items and your app is cool, then they will recommend your business to their friends and family. In no time, your sales will be up and so will your profits. If you have a small business, do not be afraid of venturing into mobile applications; they can open many new doors for success.

Meet Co-Author Irina Finkler

Irina is an electrical engineer by degree and was born and raised in what used to be the Soviet Union - more specifically, the part where she is from is Kiev, Ukraine. Her family defected to the United States in 1989 and lived in Chicago for 17 years. Irina has two boys and currently lives in Wisconsin.

Irina started her career in United States as an insurance agent, worked at Metlife for several years, and then switched to a financial company selling mutual funds. When Y2K struck and companies were desperate for computer programmers, she decided to become more technical. Irina took computer classes and went to work as a computer programmer.

Irina has always loved to work on computers and was always fascinated with how things work. Computers seemed to be the most interesting subject at the time. However, once she began working as a computer programmer, Irina started to look around at all the new &

different ways to use computers and found the internet marketing world.

She started out doing a little bit of Affiliate Marketing and then slowly but surely worked her way around to offline marketing. Irina found her niche with local marketing and the latest trend, mobile marketing. Her company now concentrates more on mobile marketing and especially on directory apps. Currently Irina is designing and developing a directory app. It works perfectly in smaller communities where Groupon or any other huge media advertising opportunities are not available.

Local Directories

Having your business information in national, regional and mobile local directories can be the difference between new customers finding you or finding your competitors instead.

The following screenshot taken recently for the keyword phrase "dentist San Diego" shows that local directories take up a majority of the organic real estate of Google's Page 1.

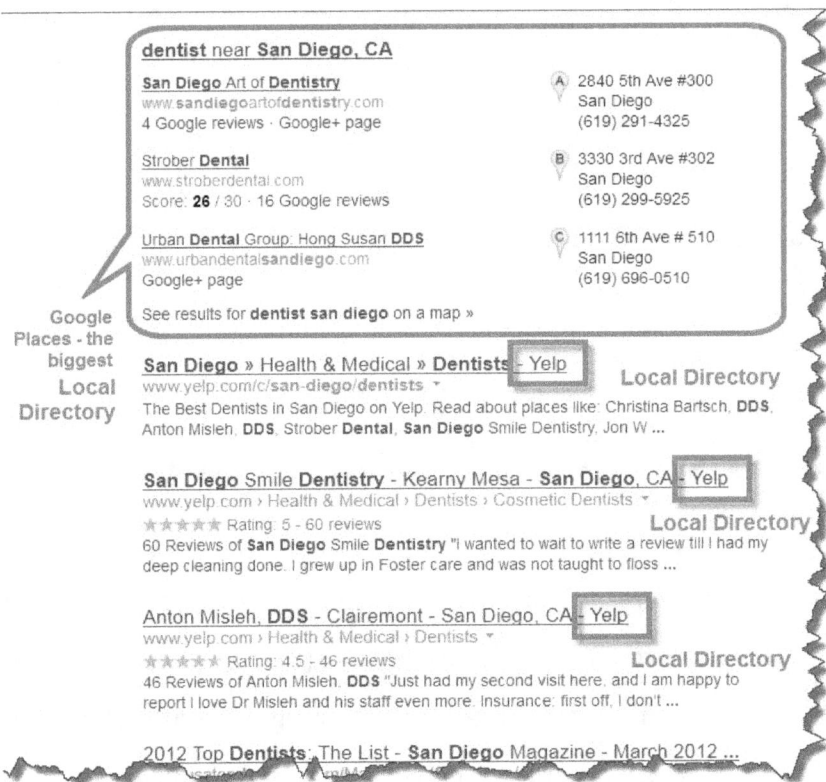

First up are the listings that appear in Google Places -- the biggest local business directory and the one that counts the most. Just look at its location on the page -- every business wants to see their information right there on Page 1 of Google.

After Google Places, of the ten organic (non-paid) spots on page 1, six of them are Local Directory listings (Yelp & YP.com), two of the listings are from a Magazine -- can't imagine that's cheap to advertise in, one is a whitepages list of dentists. I didn't count and

ONE spot on that page is held by a dentist's website ... not a single one.

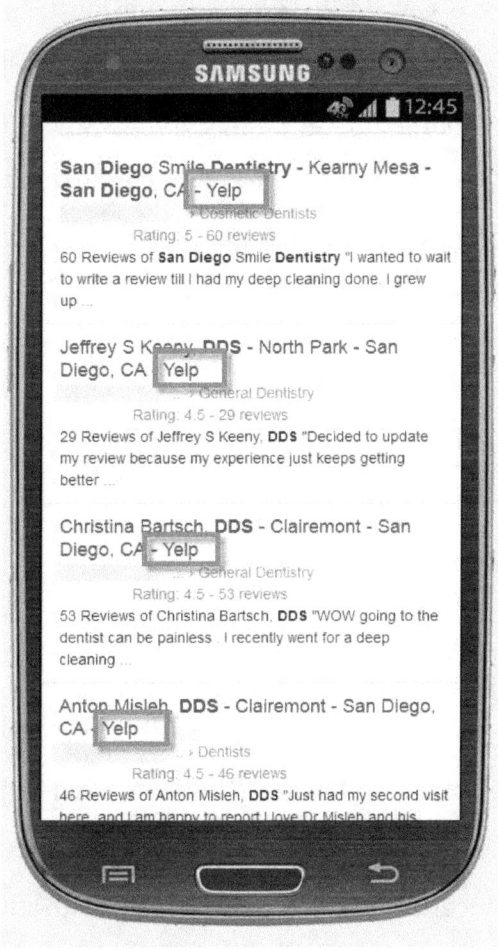

If anything should convince you that listing your business in local directories is crucial for new customers finding you via search and mobile customers finding you on their phones -- this screenshot should do it. You can do a search for any business type for any medium to large city and you'll find the same results. Local directories are everywhere.

If you pull up the same search on a mobile phone, you'll see just about the same type of results.

National Local Directories

National Local directories are generally what can be referred to as the "big box" directories, such as Yelp! & Google Places (now called Google Plus Local). Then there are the regional & city local directories (both website & mobile app versions) that are set-up and usually managed by organizations, businesses, or even individuals located in your area. These regional & city local directories can be for all business types or may only allow inclusion by a specific business-type or niche, such as dentists or home improvement companies.

There are basically four methods for creating & updating your business listing online as well as cleaning up your brand presence (if necessary) on key local directories out there.

1. Local Business Database Providers

These include Localese, Infogroup, Dunn & Bradstreet and many others. These services will push your business's information out to multiple directories. You enter your business's information once and it is propagated out to the local directories they service without you having to do anything further.

The good news is that they are relatively cheap -- anywhere from $20 to $100 per month.

The bad news is that they take up to 6 months to update your listings and the information is incomplete.

These services sell the data that you provide to the Yahoos, Yelps, and Googles of the world and the big players take these multiple feeds and add them their listings. The problem is that the big guys send out a lot of feeds and some feeds may override others. This means that sometimes old information overwrites newer updated information. Basically, these services can be unreliable if it's important that the right information gets to the right place.

Even worse, the listings are not complete -- it does not include photos, videos, or detailed information to best portray your business. What's the point of being in a local directory if your profile is so incomplete that potential customers can't find complete information about you to the point it may look like you aren't in business any longer? Consumers have gotten savvy enough to tell the difference between a "scraped" or fed listing and a well-cared-for manual business listing.

2. Centralized Local Listing Services

These are services that allow you to update your business listing directly but all from one location. Yext is such a service that comes to mind. They allow you access to all the featured of the rich profiles to put your brand in the best light. The more complete the profile, the better you generally rank on each of these directories. They want to promote the more beautiful profiles. And just as importantly you can update them at any time. The major downside to Yext is they do not include the most important directory of all – Google Places (Google Plus Local).

3. Do-It-Yourself

You can do all this work yourself by going to each local directory site and entering your business information yourself. It's time-consuming, frustrating, repetitive, & boring. But if you decide to try it yourself and you are in the US, HubSpot has a list of the Top 50 Local Business Directories that you can use to get you started. (http://blog.hubspot.com/blog/tabid/6307/bid/10322/The-Ultimate-List-50-Local-Business-Directories.aspx)

4. SEO and/or Local Internet Marketing Firm

Finally, there are providers that update your profiles directly and claim all of your profiles. This type of personalized service is usually available to you from an SEO (search engine optimization) or Local Marketing firm. This type of service will get your business set up correctly in anywhere from 50 to over 100 directories, including Google Places. Having a listing in a wide variety of directories is as important as having a profile that is completely filled out with relevant, up-to-date information -- including photos & videos. Manta, TripAdvisor and many other national local directories have millions of users. The key difference in hiring a Local Marketing firm to handle this for your business is they will make sure that your profile is set-up correctly, accurately and that your business listing is claimed.

Claimed profiles rank higher in most of the major directories. You own the profile (you can log in to each directory site) as opposed to having a data feed that will hopefully add or update your information in 6 months or so. Your presence on an increasing number of these directories becomes a marketing asset that you control. The biggest reason to go this route is that you have more time to spend running your business while experts are creating/updating your directory listing for maximum results.

Mobile Directories Drive Business to the Door!

How do you reach people when they are ready to make a decision?

Simple, your information just has to be available for them to find. Your sister is on vacation and wants to find a great seafood restaurant, who do you think she will ask?

Her phone. Your sister will search for local information on her phone, may find and download a local directory app and if your company is prepared, she will find your information.

It's a fact; people use their phone to find what they are looking for when they are ready to act.

This is the big new thing for marketing your business.

Local Mobile Directories are powerful because they work. They drive people to your door when people are on their phones. (Out shopping, vacationing, eating -- in other words, spending money!)

Mobile Directories are geared to drive new business to you.

Benefits of a Mobile Directory App

Having new customers discover your business is crucial. Many consumers will find and download Local Directory apps for their regional or city area because they don't have to wade through listings for businesses that are too far away. Mobile consumers want business information, directions, phone numbers, and services offered and they want them now. Many local business directory apps have an "In My Vicinity" type feature that brings up the businesses that are right around where they are sitting in their car or café.

These are features that are usually only available in an area-specific Mobile Directory app downloaded from the Apple App Store or Google Play Store.

These apps serve a very specific purpose in that they only contain businesses from a specific geographic area and depending on the company that created the app, may include push notifications of sales, coupons and deals as well as business information and events.

Many times you can add your business listing to these apps for free and then if you want a Featured Listing or additional promotion/advertising then there's an upgrade fee for that per month.

Having your business listed in a Local Business Mobile App is a pretty good investment because these types of apps are generally free to consumers and will usually get downloaded and installed a lot. While many apps have a short install period, these directory apps have far fewer uninstalls as it provides very specific and details information that consumers can refer to over and over again.

The benefits of mobile and local directory optimization are clear. You get:

- Mobile customers can find you easily using a Mobile Directory App or by Googling you and your categories.

- More exposure throughout the Internet when customers search for a business like yours on their sites

- Better Google Places ranking because your NAP (name, address, phone) information is consistent across the key local directories

- Better organic search engine optimization (SEO) through all the links from these sites to your website

In short, whether national, regional, city, business-specific or in a mobile app, having your business listed in local business directories is crucial to bringing your business to the attention and into the hands of new & mobile customers.

Meet Co-Author Jack Hopman:

Jack Hopman is the founder of Local Search LLC, a company he started out of passion for internet marketing.

Jack has been working his whole career in technology. After earning his bachelor degree in mechanical engineering he worked as an engine research engineer for many years. His love for computers and marketing eventually led him to the path of internet marketing. He started with specializing himself in Google AdWords. Finding his through roots he decided to venture into self-employment in 2009 and founded Local Search LLC.

Since then he has expanded his business in many other areas of online marketing like social media, Google Places, lead generation, and mobile marketing.

After connecting with many other internet marketing gurus he learned that there was a huge demand for highly specialized internet marketing tools. Finally he could combine his software experience as an engineer with his passion for marketing which lead to several highly effective software programs developed by and for internet marketers who want to be on top of their game.

In 2013 Lead Tracker Jack, Lead Finder Jack, and Ranker Tracker Jack were released. Jack is recognized by many highly appreciated internet marketers as the new man in town for developing new internet marketing tools and highly experienced AdWords marketer managing over 50 campaigns worth over $1,300,000.00 US.

Google AdWords Made Easy

Most people look for local services using Google as their search engine.

SEO (search engine optimization) is hard; organic listings change every time Google updates their algorithm as well as every time a business adds a new webpage, adds, or changes their keywords and it takes lots of time and work to get anywhere close to Google's first page.

As you know, the top 3 positions are the ones that really count in terms of traffic. Google knows that very well and made the very first 3 visible positions (usually on a pink background) on the page paid ads via their PPC platform Google AdWords.

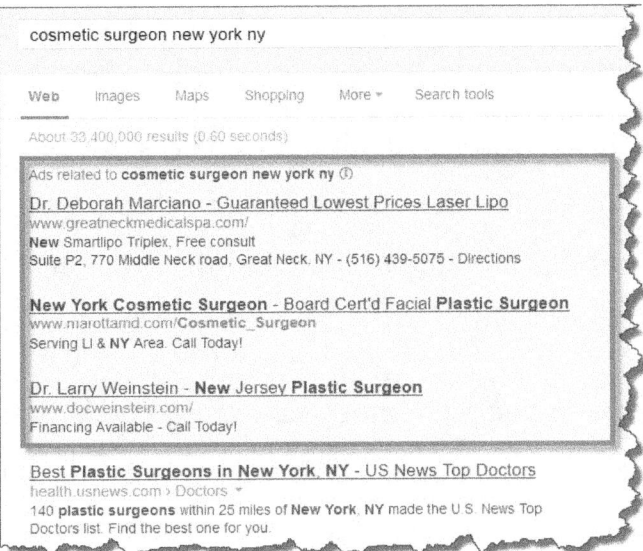

High search keywords are relatively expensive but local intent keywords are not. Below you will learn exactly which keywords to pick and how to create a good ad to generate lot of calls for your lead gen customer at minimum cost.

Choosing Keywords

A keyword is generally a word or a phrase that a person will use when they are searching for a product or service. If the search is localized to a geographic area, the searcher may include the city and state they want to localize the results for.

For example, let's take a customer who is looking for a roofer. A good keyword would be *"roofer Boca Raton" or "new roof Atlanta GA"* since it has the keyword with city name and may include the state. You want to find out what the high searched keywords are and then drill down to only those businesses that offer that service or product in your local area.

In order to find high search keywords enter your first choice keyword in google.com and scrape the top 5 AdWords URLs and the top 5 natural search or Google Places URLs.

For example:

- www.hyerfortlauderdale**roofing**.com/

- www.fortlauderdalebest**roofers**.com/

- www.atlas**roofing**fl.com/

- www.tigerteam**roofing**.com/**BocaRaton**

- www.planetconstruction.net/

- www.graboski**roofing**.com/

- www.caldwell**roofing**.com/

- www.aabco**roofingbocaraton**.com/

- www.felixsapienza**roofing**.com/

- www.royalpalm**roofing**andgutters.com/

Next, go to the Google AdWords Keyword tool:

https://adwords.google.com/o/Targeting/Explorer?__c=10000
00000&__u=1000000000&ideaRequestType=KEYWORD_IDEAS

and enter the websites one at a time from the scraped URLs you copied down and gather the highest searched keywords.

If you already have a Google AdWords account, you can log in and then find the "keyword Tool" under "Tools and Analysis". You are given more results per search and more options if you log in prior to your searches.

You want to make sure the "Country" setting is set to where your business is located. Also set the match type to "Broad" so that related keywords get shown.

Click the tab "keyword ideas" and sort by "Local Monthly Search". If you don't see Local Monthly Searches in your results, you'll need to click on Advanced Options & Filters and add a Filter Idea called "Local Monthly Searches."

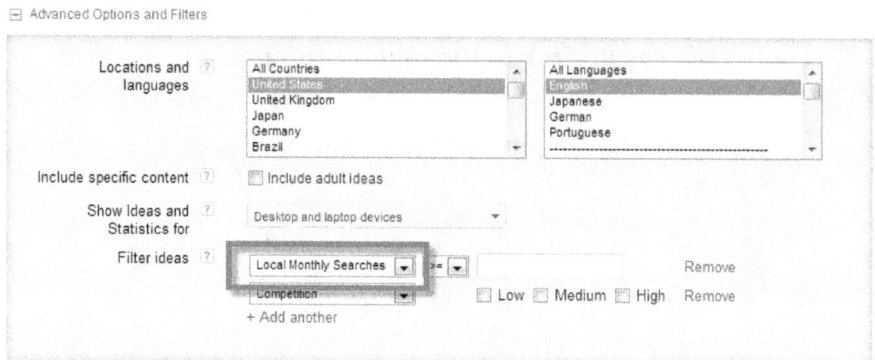

In the Keyword column select the top 10 searched keywords which represents a searcher looking directly for your service and not only for information.

Example *"about roofing"* is a high searched keyword but the visitor is looking for general information.

The keyword *"roofing in"* is a searcher which is looking for a local company and a good keyword.

For example I would select:

- Roofing in (2,740,000 local searches)

- Roofer (165,000 local searches)

- Roof repair (165,000 local searches)

- Repair roof (165,000 local searches)

- Roofers (165,000 local searches)

- Roofing contractors in (165,000 local searches)

- Contractor roofing (110,000 local searches)

- Roofing contractor (110,000 local searches)

- Roofing repair (110,000 local searches)

- Repair roofing (74,000 local searches)

Do this for all 10 URLs you copied from the AdWords and top search results and you will get a good overview of best highly searched keywords.

Next you add a local intention to the keyword, for example:

- City (Boca Raton)

- In

- At

- Near

- Location

- Places

- Company

- Companies

- Etc

There are 3 main ways that you can specify a keyword in Google AdWords:

Exact

[roofer Boca Raton] means the searcher needs to enter exact the search keyword in the search before Google will show the ad.

Phrase

"roofer Boca Raton" means the searcher needs to enter the search keyword as part of his search.

For example:

roofer in Boca Raton FL

Google will show the ad as the keyword is defined in phrase but will not show the ad if the keyword is defined as exact

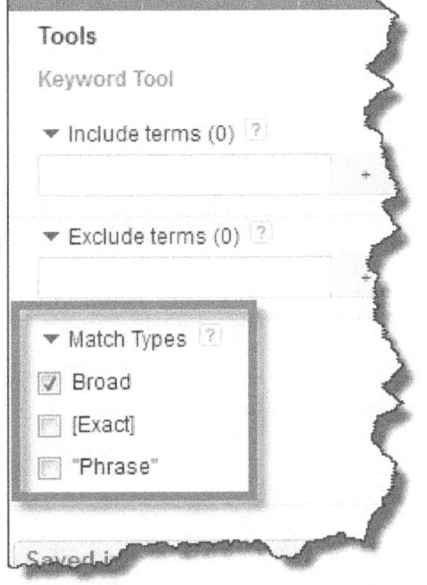

Broad

Roofer Boca Raton means the bid keyword needs to include the search keyword but the included words can be in any order.

For example:

- *Boca Raton roofer*

Google will show the ad because the bid keyword is defined as broad but will not show when it is entered as phrase.

Most of the time, Search Keywords with a local identifier can be defined as "broad."

Writing Converting Ads

A good AdWords ad answers the searcher's needs. If the searcher resonates with the ad, they will click on it to find out more.

The head line of the ad is most important, followed by the URL. The ad needs to answer the person's (keyword) need and promise the solution.

Here is the basic ad structure you should follow:

First Line: Company name Or Keyword (25 characters)

Second Line: Pain point or most important element of need

Third Line: Solution and Action

(total of 70 characters for 2nd & 3rd line combined)

Display URL: 35 characters

To come up with a good ad you need to have some knowledge about the business.

Mainly you need to know what attracts customers to that business, and what their pain point is.

A quick way to get a good ad created is to search with your high search keywords with the 10 largest cities as local identifier

For example:

Roofer New York

Roofer Los Angeles

Roofer Chicago

Roofer Houston

Roofer Philadelphia

Roofer Phoenix

Roofer San Antonio

Roofer San Diego

Roofer Dallas

Roofer San Jose

Roofer Jacksonville

Scrape the top 3 ads and pick elements out of it to compose your own ad.

For example:

Tampa Roofing Contractors

Locally Owned Roofing Company

Get A Free Estimate Today & Save!

www.ABC-Roofing.com

Injured in an Accident?
www.1800needhelp.com/
You May Be Entitled to $10,000 +
Free Case Evaluation. 24/7 Call Us

Top LA Plastic Surgeons
www.drsimoni.com/
Facelift, Nose Job & Eyelid Lift
Board Certified Specialist-Dr 90210

Make $7487 a month?

www.pochiring.com/Work

Find Out How this American Mom
Makes $7,847 a month From Home.

The above are ads that tell you exactly what you will be getting if you click those links.

Test different ads, at least 2 at the same time, then the pause ads which have the lowest CTR (click through rate).

Do not create an ad for something you don't offer just to get people to come to your site. That will just make a potential customer angry.

For example, don't promise a discount if you don't actually give a discount, or if you don't want to position the company as a low price company.

And be very careful that you are not vague or confusing in your ads. Be specific and engaging. Here are some examples of ads that didn't do so well.

Search query = used cars

Ad Copy:

Medical Equipment

Second Hand Medical Equipment

Spareparts & Tubes Extensive Databat

www.example.com

The company that put up this ad was too broad in their bid terms. They wanted only people who were searching for "used cars" but ended up bidding for any broad keyword search that included the word "used."

Search query = web hosts

Ad Copy:

So you are going to pay

top $ for exactly what we offer –

your money Ralph! Test us for free.

www.example.com

This ad completely stumps me. The keyword phrase was "web hosts" but the ad doesn't say what they sell, what they offer or who the heck Ralph is???

Targeting Locations

The Google "Target Location" setting gives you a couple of options to define the target area:

- Cities

- Zip codes

- Radius around zip codes

The size of your target area is dependent on several factors like your Google AdWords budget, your experience to create a winning campaign, how well the website convert and how many people live in that location.

I've seen the best results by defining the area within a certain radius around the business zip code. That way I can easily increase the radius once results show that the campaign is profitable.

Ad Extensions

Ad Extensions are extremely important to get high click through rates. Google allows you to show the local business address and phone number for the top 3 ads.

You can either enter those manually or link to the Google Places listing/account. I prefer to enter them manually so I have control what gets shown.

Website

Sending a visitor to a website makes sense only when the website visit results in an action. The page your ad sends customers to needs to answer what was promised in the Ad and at first glance.

- It needs to offer a compelling headline.

- It needs to show social proof, so put testimonials and social elements front and center (and above the fold) as proof of your good reputation on your website.

- Show phone number and Location in the Header and on every page.

- Have a clear and concise Call-To-Action above the fold. What do you want the visitor to do? Call you, fill out a form for a free report or estimate, etc. Don't assume they'll know what to do once they get to your site – tell them.

Thanks to Jack Mize (http://www.JackMize.com) for the website text below – it's a template of what to put on your local business website to make it very simple:

If you are thinking about repairing your roof, you probably have a lot of questions.

- *"How much can I afford?"*

- *"How much down payment do I need?"*

- *"What kind of credit do they look for?"*

- *"What kind of roof do I need?"*

At Roofer LLC we've helped hundreds of Boca Raton families get the right roof whether new or repair.

And we're confident we can help you. Even if you think a new roof or repair is out of your reach, give us a call, you may just be surprised.

To get a fast, free, no obligation estimate Call 555-555-5555 or fill out our simple roofer help form and see how Roofer LLC can help you realize the dream of your new roof today.

It is easy to edit this script for just about any local business.

Also adding an "explainer" video will help the visitor take action right away (example: http://www.labbocaraton.com)

Ninja Tricks:

- Test your ads, Test your ads, Test your ads!

- Find the best ads with keyword in other cities

- Count conversions if they surf through the website

- Look at "searched keyword" to get good info to add new keywords and negative keywords

- Test different domain names

- Did I mention that it's important for you to test your ads

Mobile PPC Strategies

According to a Google/Nielsen study released in March, (http://searchenginewatch.com/article/2255624/77-of-Mobile-Searches-Happen-at-Home-or-Work-Study) three out of four mobile searches trigger follow-up actions such as:

- further research

- a store visit

- a phone call

- a purchase

- word-of-mouth sharing

So how do you approach a mobile strategy?

- Look at the overall mobile traffic on the website from all sources. Compare that to paid search to understand how it is similar or dissimilar.

- Examine the pages per visit, time on site, and conversions from mobile (compared to desktop) to determine the engagement from mobile. If it's low, new mobile campaigns will likely show more of the same low engagement, so now is the time to optimize the website experience.

- If there is a decent conversion rate and integration with the site, consider how you can optimize that over to a more mobile experience with either a mobile site or a responsive WordPress theme, offers & coupons that will attract mobile customers looking to buy or conversion paths that are easier/quicker for users.

- Dig into paid search settings by reviewing past mobile CPCs and adjust. Review locations or time of day to identify when people are clicking your ads, trends and consider adjusting your ads to focus the maximum exposure during those times and in those locations.

- Plan ahead on ways to increase conversions based on what you know and what you will learn. Focus your optimizations on what is giving focus on bids, ad copy, landing pages, conversion path, or all of the above?

In Closing

Mobile marketing is the newest mass media with the greatest capabilities and highest marketing potential for the business community. Mobile device use is growing at a **rapid** pace and this growth reflects a growing customer base a business has the potential to tap into.

If you can imagine what it was like back when TV was infringing on radio and the explosion that resulted from that, you'll realize that mobile marketing is infringing on the internet and there is an explosion on the horizon for that.

Are you ready for the mobile explosion?

The mobile industry, as a whole, is not understood by many and is in a constant state of change. Considering the multitude of marketing tools available in mobile, as a business owner, you need to seriously consider using experts in the mobile marketing field to make the best use of your marketing dollars and time.

Don't wait till it's too late! Start making your plans to go mobile now!

Good luck – and healthy profits – with your Mobile Marketing!

Quick Start Guide

Your Mobile Marketing "Quick Start" Plan

So you're ready to take the plunge and step in the virtual world of Mobile Marketing. Here is a handy "cheat sheet" to help you make sure you're well prepared...

Your number one goal is to know your objective and also where you expect mobile marketing to fit in your overall business plan and sales funnel over the next five years.

1. My goal for mobile marketing is:

_____

_____

_____

2. Mobile Marketing can help my customers / clients most in these areas:

_____

_____

_____

3. My research on my target customers' mobile use stats and preferences has taught me that they want:

_____

_____

_____

_____

_____

4. I have identified the following areas I believe mobile marketing could bring in new customers:

_____

_____

_____

5. I have identified the following areas I believe mobile marketing could help engage the interest of my existing customer base and get them back into my sales funnel:

_____

_____

_____

6. I have decided on a mobile website and I want it to have the following:

_____

_____

_____

7. I have decided to start using SMS text messaging for my campaigns and will use it in the following ways:

_____

_____

_____

8. Service providers I have decided to use are:

	Mobile Website	SMS	Mobile App	QR Codes
Company				
URL				
Contact Name Phone/Email				
Cost/how often				
Username:				
Password:				
Other Info				

9. I plan to include the following tools in my Mobile Marketing mix:

- ☐ Mobile Website
- ☐ Mobile App
- ☐ Push Notifications
- ☐ QR codes
- ☐ Dynamic QR code marketing
- ☐ SMS Text messaging
- ☐ Virtual Business Cards
- ☐ Appointment reminders
- ☐ Scheduling Apps
- ☐ Coupons
- ☐ Contests
- ☐ Discounts
- ☐ Social media sharing buttons
- ☐ "Forward to Friends"
- ☐ Other _____
- ☐ Other _____
- ☐ Other _____
- ☐ Other _____

10. I am tracking my results!

Campaign # and Name		
Start date:		
Objective:		
End date:		
Cost:		
Results:		
Lesson learned:		
NOTES:		

Campaign # and Name		
Start date:		
Objective:		
End date:		
Cost:		
Results:		
Lesson learned:		
NOTES:		

Campaign # and Name		
Start date:		
Objective:		
End date:		
Cost:		
Results:		
Lesson learned:		
NOTES:		

Questionnaire

How do you shop?

Please circle A, B , C, D or E as your answer

1. How do you prefer to look up facts about or find the location of local businesses?
   A = on your mobile phone
   B = online
   C = in person
   D = by home telephone
   E = by regular mail

2. How do you like to search for the best deals?
   A = on your mobile phone
   B = online
   C = in person
   D = by flyers, newspapers

3. What is your preferred method of communication?
   A = on your mobile phone
   B = by e-mail
   C = in person
   D = by home telephone
   E = by regular mail

Thank you for taking time to answer! For [Your Free Incentive] and to receive updates plus VIP coupons from [Your Business], fill in your name below and include details for one or more of your favorite methods of communication.

NAME: _____
EMAIL: _____
PHONE: _____
MOBILE#: _____
ADDRESS: _____
_____

_____You have my permission to send me mobile coupons (no more than 4/6 per month). Standard data rates will apply.

Mario Brown, Brian Anderson, Irina Finkler, Susan Banks, Jack Hopman, and Jeff Cepuran

www.ingramcontent.com/pod-product-compliance
Lightning Source LLC
Chambersburg PA
CBHW051522170526
45165CB00002B/569